Impacts of Racism on White Americans In the Age of Trump

Duke W. Austin · Benjamin P. Bowser
Editors

Impacts of Racism on White Americans In the Age of Trump

palgrave
macmillan

Editors
Duke W. Austin
California State University, East Bay
Hayward, CA, USA

Benjamin P. Bowser
California State University, East Bay
Hayward, CA, USA

ISBN 978-3-030-75231-6 ISBN 978-3-030-75232-3 (eBook)
https://doi.org/10.1007/978-3-030-75232-3

This Palgrave Macmillan imprint is published by the registered company Springer Nature Switzerland AG
The registered company address is: Gewerbestrasse 11, 6330 Cham, Switzerland

To the rainbow of people for whom "Black Lives Matter", who cut through cynicism, me-ism, and hopelessness to imagine a better us, and who put their bodies in the way of racism.

To Arvie Austin Winsor, Duke Austin's late grandmother, and to all of the 675,000+ Americans killed by COVID-19 during the writing and production of this book.

Foreword

For forty years, ever since President Ronald Reagan declared that "Racism in America is Dead," Benjamin Bowser and his colleagues have been telling a different story. For White Americans, it is a startling story about the damage done to us by our own racial assumptions and policies. Now is a time when we can and must, with renewed attention, learn from the knowledge these authors bring us, for we are in the midst of a crisis; the actions and violence of White racists threaten our democracy.

In 1980 when Benjamin Bowser and Raymond Hunt published the first edition of this book, *Impacts of Racism on White Americans*, most White Americans did not imagine that racism had negative effects on their own lives. In fact, few White Americans believed racism was still an important force in American culture, damaging the social fabric in large or small ways. Hadn't the civil rights movement corrected the racial wrongs?

A common consensus among Whites was that racism, then understood as victimization or discrimination against people of color, had been reduced, and America could keep moving forward toward its national ideals. But in the late 1980s, many White people began to see through the lens of privilege that permeated life in America, conferring on Whites the advantages of being born White in a society that favors Whites. Racism was recognized by many as having an upside, which is exemption from racial discrimination. Many of us began to understand that a certain amount of power accrued to us even if we were poor. We had comparatively more help in life than people of color—in matters of respect, acceptance, pay, housing, healthcare, and legal protections—that we got because we were not being discriminated against racially. I began to suspect that these special assets might not be very good for us, and not very good for our character as individuals. I did not realize the extent to which they were actually bad for us or might be ruinous to society as a whole.

Benjamin Bowser, Raymond Hunt, and the authors for the second edition of *Impacts*, in 1996, expanded the counterintuitive evidence that privilege is a liability for White people. They collected research on the ways in which racism, in the form of White privilege, negatively impacted even privileged Whites, as well as everyone else. As authors, they collaborated with many other scholars to demonstrate yet again how racism works against Whites' wellbeing in many different areas of U.S. culture, institutions, and lives. They gathered research on the damage done by the powerful to themselves through small and large racist decisions, policies, and actions. Their diagnoses of racism's pathologies have stood the test of time.

Now Benjamin Bowser and Duke Austin have realized that these times call for a further edition of *Impacts* to reflect the unprecedented effects of racism in the Trump Era. I agree, and as I write this in February of 2021, the deadly attack on the United States Capitol carried out by violent racists illustrates more fully than ever before the "impacts of racism on White Americans."

The attack on the democracy we live within, carried out by White racists carrying Confederate flags, threatens the American governmental structure that is now intended to serve people of all colors everywhere in the nation. One impact of racism on White Americans is that it made this mob of White attackers feel entitled to force White supremacy on the democracy that is constitutionally charged with holding us all. The White racist insurrection was and is a threat to U.S. national security and therefore the security of all of us who live in and depend on this albeit incomplete democracy.

How did it come to insurrection? This is a key question the chapters in this new edition "Impacts" help to answer. The answer is that racism insinuated itself into White culture and coopted many of our minds, actions, and ideals. Though this edition of *Impacts* was for the most part written before the attack on the Capitol, still its chapters give us the authors' current understandings of how their fields of knowledge show racism being perpetuated by Whites to the detriment of themselves, afflicting even the most powerful and apparently safest members of our society. The costs to White people are documented in chapters on income, housing, health, foreign relations, gender, education, athletics, and universities' admissions policies. In most cases, what I think is being documented is not unconscious White privilege, nor conscious White supremacy, but rather what as I see as a middle ground of White control, consistently resisting the sharing of power, consistently evading the question of what might be "the greatest good for the greatest number."

Racism percolating in society led to the White mob's attack on the capitol that threatened the nation's most politically powerful White legislators, and it was White racism and violence in the temperament of Trump, the most powerful man in the United States, that incited the mob to attack the White seat of government. Their Whiteness has allowed most members of the mob to get away with their attack, suffering little punishment as of this date. But most people of all ethnicities feel damaged by the racist attack by Whites that has undermined the nation in which they, for better or worse, wish to live.

In the wake of the insurrection, this book can serve as an encouragement for White people to analyze our own psyches as well as our workplaces, schools, families, and media for the harm done by our habitual taken-for-granted racial ideas. What are racism's impacts on us? Are they good for us? Do they help us? Are they bad for us? Do they work against us? Are there some ways we can lessen dominating behaviors and attitudes that we feel are most damaging to us, our communities, and our quality of life? Are there ways we can come to understand better what we are doing?

I will tell here a story of healing that involves sharing power with those I was taught to dominate: students. I lectured; they listened. I tested; I graded. I was a strict grader and a respected teacher who "knew what she wanted." That was all right for the teaching of English usage and grammar, but when it came to the wider worlds of human experience, I joined with other college teachers from four different fields to create an American Studies curriculum that students devised for themselves. We lectured briefly for part of the time, but they taught each other most of the time, identified what they wanted to study, and told each other about what they found. They taught us. Word got out, and all of the students wanted to take our courses. Students and faculty of color could identify what they wanted to study and say. We evaded subtle hierarchies of race in the curriculum, the teaching roles, and in the whole climate of the classroom.

I feel the older system had cost all of us as teachers. The racism of the hierarchy in which White teachers set all the rules and standards paradoxically cost me, a White teacher, peace of mind. It made me feel tense, vulnerable, unprepared, on guard, never knowing enough, never good enough, exhausted by the strain of solo teaching. Though I was a very popular teacher, I felt I was never "on top of it." Now I think that being "on top of it" was a racially based ideal of superiority—knowing more than all the students about everything. I think the feeling that I must be superior to them was based on subtle training in social and racial life; I had to be able to look down on others. Did I think training in White "excellence" was detrimental to me? It got me "ahead" in White structures, but now I think in many ways it was indeed detrimental to my growth and development. It was training in keeping racialized Others out—keeping out unlicensed Others' thoughts, ideas, knowledge, cultures, and humanity. So the classroom was an excluding place.

For this reason, I understand every chapter of this book as an invitation to look at how my deeper self-interest as a White person may be served by getting smarter about structures of dominance in and around me. White certainty about who and what counts in education had cost me heavily. The freedom to connect multiculturally and across lines of age is emancipation.

Socially aware teachers who use chapters from this book know that learning about racism can make Whites who are new to the subject feel confused, defensive, offended, angry, indignant, resistant, and fearful that if they share racial power, they will lose the lives they have become accustomed to. The authors of this book are gentle with such new readers. Instructors may nevertheless see some distress in White readers who are unprepared to imagine that racism hurts even them. This is not a reason for educators to avoid the subject. Many White people who feel invulnerable now will later have to weather times when they realize the racial systems were not good for them either, and they will want to know how to begin acting on that knowledge. The chapters in this book will support them to change without blaming them for systems they did not invent.

I have a debt of gratitude to these editors who, decade after decade, carefully and persistently provided us with scholarship confirming that a racist society is not good for anyone. I feel indebted also to these many authors who have opened up the question of how the dominants may pay a high price for dominance. May we have a fourth edition that contains better news.

Peggy McIntosh, Ph.D.
Founder and Senior Associate, National SEED Project on Inclusive Curriculum (Seeking Educational Equity and Diversity)
Wellesley College
Wellesley, MA, USA

Peggy McIntosh is a Senior Research Associate of the Wellesley Centers for Women, Wellesley College. Dr. McIntosh is the Founder of the National SEED Project on Inclusive Curriculum (Seeking Educational Equity and Diversity), which she co-directed with Emily Style for its first 30 years. She is the author of the influential work, "White Privilege: Unpacking the Invisible Knapsack" (1989). Sixteen of her works on privilege, education, and psychology may be found in the book of her essays *On Privilege, Fraudulence, and Teaching as Learning: Selected Essays 1981–2019* (Routledge, 2019).

Preface

This is not just another book about racism. We present a novel way to look at racism. Instead of focusing on racism's effects on African Americans and other people of color, we explore the flip side of the race question. Does racism have any effects on White Americans? We think the answer to this question may be one of the most important unaddressed issues in the United States today, and is closely related to the January 6th, 2021, "patriots" attack on the Capitol in Washington. Neither the question nor its importance takes anything away from the extensive and well-documented literature on the damaging effects of racism on people of color in the United States. The question we ask opens the door to another side of racism's cancerous effects.

Another door opens when we explore racism's impacts on White Americans. The solution to the racial inequality and conflict is not in any of the places we have been looking: If the few racist Whites disappeared, racism would still exist. If more Blacks and Latinos were better educated and got better jobs, racism would still take its toll. If every Black person in America received reparations, racism would still exist. Empty the prisons and make all the former inmates choir boys, the police would still fear Black men and target them disproportionately. Eliminate poverty, and there would still be racial inequality. African, Latinx, Native, Asian or any other people of color are not the cause of racism. And if they are not the cause, there is nothing that can be done to them nor is there anything they can do, say, or become to end racism. The door we open is to the realization that the cause of racial inequality, disparities, discrimination, and fear is White elite's exercise of power combined with the long cultural history and endorsement of rank-and-file Whites that facilitates their power—racism. Whites are the cause as well as the solution. Until this point is reflected on, realized, and taken to heart, there may be change, but there will be little progress in defusing the race bomb and its periodic explosion.

The Need for a New Edition

This is the third edition of a classic work, *Impacts of Racism on White Americans* (1981, 1996). When the first edition was published, Ronald Reagan became President of the United States. He immediately declared racism was dead. His view was that the Civil Rights, Voting Rights Acts, and the prior decade of affirmative action had successfully dealt with racism and the nation had enough reforms. White people were no longer responsible for racial inequality; from then on, Black people were responsible for any continuing racial inequality. The first edition had two theses. First, racism in the United States is fundamentally a problem of White power, identity, and culture, unaddressed since this nation's colonial times. Second, the continuing exercise of power, and racism have substantial adverse effects on many, if not most, White Americans.

In the second edition of *Impacts* (1996), we explored the causes of White racism and its effects on Whites before there was an openness to do so. We had to convince readers that racism existed, was caused by more than a few bad apples, and had an institutional reality. A consensus formed that the remaining White racial animus was symbolic, covert, unconscious, and colorblind. The centrality of Whites in racism and its effects on them was overlooked again. Even studies of racial discrimination stopped finding overt racial discrimination. Racial discrimination became only what Blacks perceived it to be. Black perception of racial discrimination was due to their paranoia rather than real discrimination. The Supreme Court stepped in and dispelled any further questions that racial discrimination existed. It no longer existed "de facto." It existed only if one was able to prove racially discriminatory intent on the part of alleged discriminators. In effect, a robbery victim lying in front of you was not there unless you knew who the robbers were.

Then, Barak Obama was elected President of the United States. His election was held as decisive proof that racism no longer existed. How could racism still be a potent force in everyday American life when millions of White voters elected a Black President? Our central thesis that White Americans should be the focus for serious study was overlooked again. In 2016, along came the surprise election of Donald Trump as the 45th President of the United States. Then, before all our eyes, the White racial animus held by a large portion of the American public, that we had to convince earlier readers still existed, became visible for all to see. With racism as overt as it is today, it was time to re-examine the impacts of racism on White Americans.

About This Book

This book is intended to be read by undergraduates, sociologists, psychologists, social historians, foundation and government program officers, graduate students, health specialists, students of American Studies, a variety of faculty, journalists, social workers, radio broadcasters, diversity trainers, policy institute analysts, social justice activists, and the general public. We hope to reach a wide audience. Upper division undergraduates can explore how a historically unexamined belief in racial supremacy and the exercise of power through institutions and individual behaviors can produce social inequalities for Americans of all colors. The 15 chapters make the book easy to adopt in a 15-week semester. Of course, individual chapters can serve as supplement readings.

Hayward, CA, USA

Duke W. Austin
Benjamin P. Bowser

Acknowledgments

Duke W. Austin: I want to thank the teachers, professors, and mentors who taught me, who believed in me, and who encouraged me. For their outstanding teaching mentorship in my youth, I want to thank Mrs. Laughter, Eddie Edwards, and Tanya Miller. For those who taught and mentored me in college and young adulthood, I thank the late G. Richard Thayer, Stephen Whitfield, and Robert Jensen. For believing in me and encouraging me to embark on a second Ph.D. dissertation after the first one was scrapped, I want to thank Thomas Mayer, Jason Boardman, and Liam Downey. For his mentorship and guidance during a post-doc, I want to thank Elijah Anderson. For inviting me to be a part of this project and for promoting me to first editor, I want to thank Benjamin Bowser. Finally, but most importantly, I want to thank my mother, Linda Austin, and my life partner, Julie Kessler. Thank you, mom, for supporting me on every single adventure on which I have ever embarked, and thank you, Julie, for your unwavering support and for your dedication that we raise our children to be anti-racist and anti-sexist.

Benjamin P. Bowser: I owe debts of gratitude to scholars who have passed. Raymond Hunt, Professor Emeritus at the University at Buffalo, and Hubert Locke, former Dean of University of Washington Graduate School of Public Affairs. They realized that racism is a White problem and costs White as well as Black Americans. They invited me to join them in producing the first edition of this project. The late Robin Williams, Jr., of Cornell University, brought social structure to life and made it possible for me to appreciate Hunt and Locke's invitation. Many thanks to two other mentors. The late Loften Mitchell, Broadway playwriter extraordinary, humored and brought me home to Harlem. Through the late historian, Herbert Aptheker, I met the spirit of W.E.B. Du Bois. Finally, there is my brother in mind and spirit who left us way too early, Walter Stafford, of New York University. Three other contributors to this project are very much alive. Thanks to Nathalia Bowser, mother of the family, and to Terry Jones for bringing me to California State University East Bay. Special thanks to K. Deborah Whittle, the love of my life and wife, who made it possible for me to enjoy who I am and to write all that I have.

Contents

Notes on Contributors

Duke W. Austin is an Associate Professor of Sociology and the Department Chair at California State University, East Bay. Dr. Austin's teaching and research focus on race, gender, immigration, environmental justice, community engagement, and social justice pedagogy. He is a co-editor of the volume *Bringing Fieldwork Back In: Contemporary Urban Ethnographic Research* (Sage, 2012).

Jorge Ballesteros is a fourth-year Doctoral Candidate at Arizona State University pursuing his Ph.D. in Counseling Psychology. Before attending ASU, Jorge earned his M.A. in Counseling and M.S. in Kinesiology with an emphasis in Sport & Exercise Psychology from Ball State University in 2017. He is interested in studying athletic and student identity among racially diverse collegiate athletes, as well mental health-seeking behaviors among athletes. He wishes to continue developing his skills as a social justice advocate and implementing a multicultural perspective in his work with student-athletes.

Benjamin P. Bowser is an Emeritus Professor of Sociology at California State University, East Bay. A sociologist specializing in research methods and public health who focuses on African American communities and evaluations research, Dr. Bowser's most recent books are *Racial Inequality in New York City Since 1965* (SUNY Press, 2019) and *The Abandoned Mission in Public Higher Education: The Case of the California State University* (Taylor & Francis, 2017).

Scott N. Brooks is the Director of Research at the Global Sport Institute and Interim Director of the School of Social and Family Dynamics at Arizona State University. A sociologist who studies the athlete experience, Dr. Brooks has consulted for the NFL, NCAA, US Olympic and Paralympic Committee, USA basketball, MLB, Adidas, college and high school coaches and athletes. He has published in academic journals, edited volumes, and textbooks, been quoted and reviewed by the *Wall Street Journal*, *New York Times*, *Washington Post*, *Der Speigel*, and *SLAM magazine*, and been invited to speak on the topic of sport internationally. He is the author of *Black Men Can't Shoot* (University of Chicago, 2009), and is currently working on two books—one that investigates athletes' playing time, and another that studies coaching methods.

D. Anthony Clark is a Senior Lecturer at Arizona State University where he teaches interdisciplinary humanities and veterans studies. He is a racial justice scholar with 21 peer-reviewed articles and refereed book chapters and an additional 25 academic essays and reviews. He is past president of the Mid-America American Studies Association and, currently, consulting editor for the *Journal of Diversity in Higher Education.*

Chris Danielson is a Professor of History at Montana Technological University. Dr. Danielson graduated with a Ph.D. in American history from the University of Mississippi, where he specialized in the civil rights movement. A native of Houston, he received his Bachelor's and Master's Degrees in history from the University of Houston. He is the Author of two books, *After Freedom Summer: How Race Realigned Mississippi Politics, 1965–1986* (University Press of Florida, 2011), and *The Color of Politics: Racism in the American Political Arena Today* (ABC-CLIO, 2013).

Charlotte Chorn Dunham is an Emeritus Professor of Sociology. Dr. Dunham retired from Texas Tech University after 30 years on the faculty. While there, she served for eight years as Director of Women's and Gender Studies and one term as Chair of the President's Gender Equity Council. She was a Co-principal Investigator on the TTU ADVANCE-PAID Grant (2005–2009) and is currently working as Internal Evaluator for the TTU NSF ADVANCE-ADAPT grant. Her most recent research focuses on older adults in the Cape Area of South Africa, including publications on gender and decision-making of older couples and depression among caregivers for orphaned grandchildren.

Robert Fantina is an Author and Journalist, working for peace and social justice. He writes extensively about U.S. foreign policy and the Middle East, with a focus on Palestine. He serves on the boards of Canadian Voices for

Palestinian Rights (CVPR), and Canadians for Peace and Justice in Kashmir (CPJK). His most recent book is *Propaganda, Lies and False Flags: How the U.S. Justifies its Wars* (Red Pill Press, 2020). His next book, *Modern-Day Settler Colonialism: Israel & Palestine; India & Kashmir* will be available in 2022.

Stacey M. Flores is currently pursuing a Ph.D. in Sociology at Arizona State University. Flores holds a Bachelor's Degree in English, a Master's Degree in higher and postsecondary education, and a Master's in counseling. She teaches sociology, college readiness, family and cultural diversity courses. Her research is focused on the intersections of race, sport, and gender. She has worked on projects for the NFL, the US Olympic and Paralympic Committee, New England Women's and Men's Athletic Conference, and high school coaches and athletes.

Kalemba Kizito is a Visiting Assistant Professor at the University of Denver. His research agenda focuses on the border in all its mediated forms as a transnational and global apparatus of biopolitics. Grounded in African and Afro-diasporic identity, his work moves the conversation on borders away from its Euro-Western and U.S. situated frames, to theorize how the cultural politics of borders functions globally with emphasis on their connections to African and Afro-diaspora studies.

Patricia A. Maloney is an Associate Professor of Sociology at Texas Tech University where she is the Director of Community and Urban Studies. Dr. Maloney specializes in the sociology of education, focusing primarily on how moral choices are made in schools. She has a special interest in assessment, sociological theory, and qualitative methods.

Jonathan M. Metzl is the Frederick B. Rentschler II Professor of Sociology and Psychiatry, and the Director of the Department of Medicine, Health, and Society at Vanderbilt University in Nashville, Tennessee. He received an M.D. from the University of Missouri, an M.A. from Stanford University, and a Ph.D. in American Culture from the University of Michigan. Winner of the 2020 Robert F. Kennedy Human Rights Award, the 2020 APA Benjamin Rush Award for Scholarship, and a 2008 Guggenheim fellowship, Dr. Metzl has written extensively for medical, psychiatric, and popular publications about some of the most urgent hot-button issues facing America and the world. His books include *The Protest Psychosis, Prozac on the Couch, Against Health: How Health Became the New Morality* (Duke University Press, 2003), and *Dying of Whiteness: How the Politics of Racial Resentment is Killing America's Heartland* (Hachette UK, 2019).

Pamela Oliver is an Emeritus Professor of Sociology at the University of Wisconsin Madison. For many years she taught a course on racial-ethnic movements and politics in the United States. Dr. Oliver's current NSF-funded research is on news coverage of Black protests in mainstream and Black newspapers 1990–2020. She also does public sociology work on racial disparities in criminal justice in collaboration with community activists.

Fred L. Pincus is an Emeritus Professor of Sociology at the University of Maryland Baltimore County. Dr. Pincus is the Author of several books including *Reverse Discrimination: Dismantling the Myth* (Lynne Rienner Publishers, 2003), *Understanding Diversity: An Introduction, 3rd Edition* (Lynne Rienner Publishers, 2021), and *Confessions of a Radical Academic: A Memoir* (Adelaide Books, 2020).

Michael Reich is a Professor of Economics at the University of California, Berkeley and co-Chair of its Center on Wage and Employment Dynamics (CWED). He served as Director of Berkeley's Institute for Research on Labor and Employment from 2004 to 2015. Dr. Reich received a Ph.D. in Economics from Harvard. His research publications cover numerous areas of labor economics and political economy, including the economics of racial inequality, the analysis of labor market segmentation, historical stages in U.S. labor markets and social structures of accumulation, high-performance workplaces, union-management cooperation, Japanese labor-management systems, living wages, and minimum wages.

Jacob S. Rugh is an Associate Professor in the Department of Sociology at Brigham Young University. Dr. Rugh's research on housing discrimination and immigration has been published in leading journals and cited in several federal civil rights cases that have secured justice and defended immigrant rights. His latest article, "From Crisis to Progress: Housing and Latino Youth since 2000," appears in the July 2021 issue of *The ANNALS of the American Academy of Political and Social Science.*

Lisa B. Spanierman is Professor and Head of the Faculty of Counseling and Counseling Psychology at Arizona State University. Dr. Spanierman's research focuses on White individuals' racial attitudes, racial justice allies, and racial microaggressions. She has co-authored more than 70 articles and chapters, an edited volume, *Unveiling Whiteness in the 21st Century: Global Manifestations, Transdisciplinary Interventions*, (Lexington Books, 2015), and a book, *Microaggressions in Everyday Life, second edition,* (Wiley, 2020). She is a Fellow of the American Psychological Association.

List of Figures

List of Tables

1

Introduction: Impacts of Racism on White Americans in the Age of Trump

Duke W. Austin and Benjamin P. Bowser

Abstract If asked to define racism, most people describe a dislike of people from other races. While that approximates a definition for personal prejudice, it fails to account for the full reality of racism. Racism, instead, is a system of power, characterized by a racial hierarchy. The fact that the popular definition fails to account for power or hierarchy is no accident. In fact, it speaks to the reason that so many White Americans can simultaneously reject racism as popularly defined, engage in racist behavior without knowing it, benefit from White privilege, and suffer from the impacts of racism. In this introductory chapter, we explore how a complete definition of racism helps establish the four primary theses of the book. First, we assert that White Americans are the source of racism. White Americans are the inventors, propagators, and benefactors of racism. Second, we argue that White Americans also suffer a cost for being the source of racism. White Americans gain unearned privileges from racism relative to people of color. Third, the costs of racism are not equally spread among all Whites. White privilege intersects with other social identities such as gender and socioeconomic class. Fourth, modern elites use

D. W. Austin · B. P. Bowser (✉)
Department of Sociology, California State University, Hayward, CA, USA
e-mail: benjamin.bowser@csueastbay.edu

D. W. Austin
e-mail: duke.austin@csueastbay.edu

D. W. Austin and B. P. Bowser (eds.), *Impacts of Racism on White Americans In the Age of Trump*, https://doi.org/10.1007/978-3-030-75232-3_1

social institutions to enforce the racial hierarchy in order to achieve their own goals. This has been especially true during the age of Trump. Donald Trump exploited existing racial divisions to garner White support in cutting taxes for billionaires, gutting social safety nets for those on the margins, and fortifying his own claim to power. In the process, everyday Whites suffered the impacts of racism.

Keywords Race · Racism · Impact of racism on Whites · White identity · Whiteness · Donald Trump

We know surprisingly little about how White Americans experience racism. In the voluminous writing and commenting about racial and ethnic groups, most of the attention focuses on Black, Indigenous, and other People of Color, known as BIPOC. For most BIPOC, racial prejudice and discrimination are barriers to equality, and the cause is racism. Often times, however, it is not even clear that White Americans are the source of prejudice, discrimination, and racism. This book explores one of the least known aspects of life for Whites: the impact of racism on White Americans. We know that Whites experience White privilege, but we ask if they also pay a price for being the dominant racial group? In other words, does racism against BIPOC have any negative costs on Whites themselves, and are those costs shared equally by all Whites? In answering these questions, we advance four primary theses:

First, we assert that White Americans are the source of racism. This is not a book about the mythical "reverse racism" that the title may lead some people to believe. Reverse racism, which would indicate that Whites are now on the bottom of the racial hierarchy, is not real. Racism, as commonly understood with Whites at the top of the racial hierarchy, is real. White Americans are the inventors, propagators, and benefactors of racism. White Americans continue to dominate every American institution—including government, economy, education, healthcare, housing, media, and so on. Racial inequality is real, and BIPOC people do not oppress themselves.[1]

Second, we argue that White Americans also suffer a cost for being the source of racism. White Americans gain unearned privileges from racism relative to people of color. For instance, poor White Americans tend to receive more privilege and experience less oppression than poor Indigenous Americans, and middle-class White women tend to receive more privileges and

[1] BIPOC people do not oppress themselves except perhaps when individual people of color side with the dominant White race to oppress other people of color for their own benefit, as Ibram Kendi (2019) observes.

experience less oppression than middle-class Black women. We do not contest these claims. However, White Americans also pay a price for racism. For example, overall White income is lower in racially unequal counties and higher in racially equal counties (see Reich in this edition). In addition, White patients suffer more when BIPOC people are excluded from healthcare than when healthcare is universally applied (see Metzl in this edition).

Third, the costs of racism are not equally spread among all Whites. White privilege intersects with other social identities. For example, elite Whites benefit more from racism than poor and working class Whites, while poor and working class Whites bear more of the costs of racism (see Fantina and Danielson in this edition). Similarly, White men benefit more from the racism applied to women of color, and White women bear more of its costs (see Dunham this edition).

Fourth, modern elites use social institutions to enforce the racial hierarchy in order to achieve their own goals. This has been especially true during the age of Trump, which roughly coincides with his time in office (2017–2021) but also has roots in the American Tea Party and will almost certainly have repercussions for years to come. Donald Trump exploited existing racial divisions to garner White support in cutting taxes for billionaires, gutting social safety nets for those on the margins, and fortifying his own claim to power. He cozied up to overt White supremacists when he said there "were very fine people on both sides" of the violent Charlottesville protests, and he stoked racial divisions when he called Black Lives Matter protestors "thugs." When Trump needed an army to support his claim to power, he told the Proud Boys to "stand back and stand by." In the end, Trump kept his wealth, built his political influence, and twice avoided being convicted of impeachment. Meanwhile, Trump's mostly White supporters ended his presidency in greater economic turmoil than when he assumed office, and Trump failed to pardon even a single supporter who stormed the capital on his behalf on January 6, 2021.

So, as an orientation to the central chapters of this book, we need to better understand the definitions of racism and race, White identity and Whiteness, and the age of Trump, and we need to understand how racism is deployed through institutions. Then, we will take this knowledge to explore the impacts of racism on White Americans in the age of Trump.[2]

[2] We capitalize Black, Brown, Asian, Latinx, Indigenous, and White when referring to races, cultures, and ethnicities. We capitalize White along with all other racial-ethnic groups to avoid implying that White people and Whiteness are a given or that they are solely definitive of American culture. To capitalize White is also grammatically consistent in usage with other ethnic-racial terms. In addition, capitalization of White is essential to invite White people to acknowledge and understand White supremacy within Whiteness and engage in efforts to dismantle it. Our usage is consistent with

The Definition of Racism and Race

In this introduction, we want to define racism and provide the reader with essential background to gain the most from the following chapters. A popular definition of racism states that racism is "a belief that race is a fundamental determinant of human traits and capacities and that racial differences produce an inherent superiority of a particular race" (Dictionary 2021). However, we find this popular, dictionary definition woefully inadequate. It states that racism is a belief, and that implies that racism only exists when the belief is present. However, Robert Merton (1949) showed decades ago that racism can occur even when no one believes in racial superiority. Merton argued that the "unprejudiced discriminator" is someone who, despite their own freedom from racists beliefs, supports racial discrimination when it is easier or more profitable to do so. For example, the employer who refuses to hire people of color for fear of losing customers is still engaged in racist behavior even if the employer does not hold the racist beliefs themselves (Merton 1949).

More importantly, the popular, dictionary definition fails to account for the power structure and the racial hierarchy that is inherent in racism. By failing to include an analysis of social power or racial hierarchies, this definition puts all races on equal footing. In other words, any person from any race can believe that their race is superior, which allows any race to be the cause or recipient of racism. We assert, however, that racism does not happen in a power vacuum and not all races have equal footing. Imagine, for example, the origin of racism: slavery. The White slave master and the Black slave may have both believed that their own race was superior. Whereas the White slave master's belief had profound impacts on the Black slave's social status, the reverse was not true. The White slave master could kill a Black slave with impunity, and his belief in White superiority would justify the act. On the other hand, the Black slave's belief was unlikely to have any impact on the White slave master's power or status. Even during violent slave uprisings in which the White slave masters were killed, the White slave masters died free and the Black slaves remained slaves, hunted by society for their transgressions. Slavery has been outlawed in the United States for over 150 years, but the power structure that put Whites at the top of the racial hierarchy lives on. White Americans continue to dominate every American institution—including government, economy, education, healthcare, housing, media, and so on. As such, racism must be viewed within that racial hierarchy.

---

the recommendations of the MacArthur Foundation and The Center for the Study of Social Policy, among others.

Racism, instead, is a system of power, characterized by a racial hierarchy.[3] Racism grants unearned dominance and advantages, known as privileges, to the dominant race and oppresses the subordinate races. Racial discrimination is racism in practice or in action. Racial discrimination occurs when members of the dominant race receive unearned privileges and when members of subordinate races are oppressed. Racial prejudice, the belief or ideology that one race is superior to other races, is a product of racism, not a cause. Prejudice, the belief or ideology, serves to justify and enforce the racial hierarchy.

In almost all postcolonial societies, including and perhaps especially the United States, Whites are at the top of the racial hierarchy, meaning they are the dominant race. For this reason, we assert that Whites are the source of racism. Members of any racial group can hold prejudiced views about their own or other races. However, the system of power that is racism only works in one direction, from the top of the racial hierarchy toward the bottom. In other words, members of the dominant race are never the recipients of institutionally-derived racial discrimination, only the perpetrators. Furthermore, only the members of subordinate races experience racial discrimination even though they can perpetuate it against members of their own race and against members of the races below them in the racial hierarchy.

Racism exists in three primary societal domains—institutions, cultures, and individuals. First, institutional racism describes when the institutional system of power is designed to benefit the dominant race at the detriment of subordinate races. Whites dominate the power structure of the overwhelming majority of institutions in the United States, including government, healthcare, education, and businesses, to name a few, and Whites tend to benefit the most from those institutions. Second, cultural racism occurs when a culture is biased toward the cultural norms, mores, traditions, and practices of the dominant race. The fact that middle-class White norms are the default, unchallenged norms for the United States is an example of cultural racism. Third, individual racism is when individual people feel, think, say, and do things premised on cultural racism. Examples are a White student being surprised that a Latino student is good in math and a greeter who seats Blacks in the back of the restaurant near the kitchen to not offend White diners. Individual racism is also demonstrated by consciously or unconsciously doing things to produce institutional outcomes favorable to the dominant race and disadvantageous to subordinate race members. Again, examples are local

---

[3] This definition of racism is an amalgamation from numerous activists and scholars that precede us. For further reading, we suggest Robert Merton (1949), Herbert Blumer (1958), Peggy McIntosh (1988), Michael Omi and Howard Winant (2014), and Eduardo Bonilla-Silva (2018).

White political leaders who gerrymander political districts to keep people of color from having their own political representation or county supervisors who base school funding on property taxes.

Like racism, the popular definition of race is also incorrect. Most people believe that race is a genetic or natural category, largely defined by skin color. However, *race itself is a social construct, created primarily by those with social power to maintain a hierarchy of power* (Omi and Winant 2014). That means racial categories—such as Asian, Black, Indigenous, White, and others—are human inventions, not outcomes of nature. It means races cannot be determined by biology or any other natural science. It means that natural science cannot accurately distinguish one race from another. Races exist only because the socially-constructed racial hierarchy, with Whites at the top, has been embedded in the very structure of society. Race, therefore, is not benign. It has been described as our most dangerous myth because it is used by people in power to arbitrarily divide humans into hierarchies from superior to inferior.

Who invented race? Wealthy English planters who settled in early Virginia and North Carolina after 1650 invented the American racial classification system. Of course, they put themselves at the top of the hierarchy, and they put the enslaved Africans and Native Americans in their possession at the bottom.

"White people" did not exist in North America before the sales of the first Africans in 1619 to colonists in Jamestown, Virginia. In early colonial America, the elite planter class was intent on building wealth and power. The more laborers that they could get, the more land they could work, and the more wealth and power they could generate. Nevertheless, they had two problems. First, they could not force enough Native Americans into bondage or persuade enough impoverished English, Scots, Irish, and other Europeans into indentured servitude. Native Americans could escape into uncolonized territory and rejoin their tribes. European immigrants were reluctant to become indentured servants because it meant they had to surrender their personal freedoms and do demeaning, dirty, and often dangerous work during the most productive years of their lives. Second, planters faced potential rebellion. The overwhelming majority of the population consisted of laborers, and the laborers were growing more and more resentful of the planters. Planters' days were numbered if they did not find a way to defuse the growing danger. The 1676 Beacon Rebellion in Virginia was a violent signal of this danger.

The solution to both problems, inspired by security measures pioneered in the British Caribbean, was to import progressively larger numbers of Africans to do the planters' work and generate wealth. Simultaneously, they passed laws that ended indenture contracts for Europeans and elevated them to the status of "free labor," and reduced Africans into permanent bondage—slavery. The planters' goal was to unite European immigrants and their descendants with the elite planter class to create a "buffer" between the elite planter class and the slaves themselves.

In other words, the elite planter class invented the concept of "White people" to replace multiple, often hostile, and competing European ethnic groups with a newly constructed master race while simultaneously distinguishing them from enslaved Africans and Native Americans. This was a uniquely American amalgamation of peoples and color-coded concepts of race. "Free White people" were a step above enslaved Africans and Native Americans. To ensure their new privilege of freedom, "White people" protected the elite planter class from slave revolts and also kept the increasing numbers of impoverished White laborers in line. White people, especially the poor, served as militia to control slaves, put down slave rebellions, and capture runaways. Poor Whites avoided protesting their impoverished conditions because freedom itself was on the line. Unless they supported the elite planter class and adopted the new racial hierarchy, poor Whites could risk being returned to indentured servitude.

The elite planter class made the concept of race consequential over the next two hundred years by passing laws to enforce the racial hierarchy and punish violations of that hierarchy. When slavery ended in 1863, Jim Crow segregation soon emerged after Black Reconstruction to continue enforcing the racial hierarchy. When Jim Crow segregation ended with the Civil Rights Movements of the 1950s and 1960s, the hierarchy continued more covertly in U.S. culture, institutions, and individuals. Racism, therefore, was intrinsic to creating the White race, enforcing the racial hierarchy, and serving as the central rationale and defense for slavery. Whites were the superior race, and Blacks (called then Negroes), Native Americans, and other peoples of color were inferior races. The creation of this belief and defense of slavery was more than the birth of the White race—it was the birth of White identity and of White culture, otherwise known as Whiteness.

White Identity and Whiteness

White identity, in one understanding, is the simple act of identifying as a member of the socially constructed White race. Where one fits in the racial hierarchy is prescribed by society. So, identity as "White" is prescribed—it is initially not a matter of choice. In another words, "White identity" is the degree to which one identifies with and engages in *Whiteness*. *Whiteness* refers to White culture. But, here, there is a question in search of answers and resolution. We know that White supremacy is part of White culture. Does Whiteness have cultural content that is more than White supremacy? If so, what is it? We believe that the anti-racist White identity rejects White supremacy and builds upon allyship with people of color. To better understand the anti-racist White identity, we must better understand White identity and Whiteness itself. In the following paragraphs, we briefly outline studies of White identity and Whiteness and discuss their connection to Donald Trump and White identity politics.

Starting with her 1982 dissertation, *White Identity Development*, Rita Hardiman examined how White Americans develop a sense of racial identity in what is essentially a White supremacist society (Hardiman 1982). One would hope that White identity and White supremacy are separate entities, but what she finds is an intimate connection between them. White identity is a product of White supremacy and they are inseparable. A theme in writing on Whiteness is that it maintains form and integrity in opposition to Blacks and Native Americans (Omi and Winant 2014). White identity was created to distinguish people of European ancestry from Others and continues to serve that purpose. European immigrants established themselves not only as a working class, but also as White and free by distancing themselves from any comparison to slaves (Roediger 1991). If people of color disappeared, there would be no need for Whiteness and race supremacy.

In 1988, Peggy McIntosh wrote autobiographically about 46 ways in which she saw she had White skin privilege when contrasting her experiences with those of her African American colleagues in the same building and line of work (1988). Her work emphasizes that these White skin privileges, or advantages, are unearned, conferred by birth and physical appearance in a racist society that favors Whites. Thandeka (2020) argues that Whites experience "White shame" when learning to be White; this is an update of McIntosh's insight. The novelist Toni Morrison repeats both points: White supremacy is intrinsic to Whiteness and Whiteness is dependent on an oppositional Other (Morrison 1992). However, as part of supremacy, Whiteness also must be considered normative, unbiased, mute, and neutral—not just White, but male as well. Morrison's point about Whiteness being male is not just a literary device. Matthew Jacobson agrees with Morrison. He asserts that Whiteness is specifically a product of very aggressive male activities: conquest, enslavement, imperialism, and segregation (1999). Lopez highlights the same point—that Whiteness came out of European colonialism and conquest (2005). This is not just a point of history; it is a belief in the right to exploit others and perhaps even the need to do so is also intrinsic to

Whiteness. Whiteness, according to Lopez, is about mastery over people and things.

Ashley Jardina, in *White Identity Politics* (2019), holds that the White identity we see on display with Trump's loyal followers does not necessarily have deep roots in history. White Americans never had to think about race or their identity until recently. Their prospects of becoming a minority have prompted for many a need for racial solidarity, consciousness of Whiteness, urgency to preserve their privileges, and to do so through politics—White identity politics. Furthermore, some with a strong sense of White identity are overtly prejudiced, while others are not. Others, still without a strong White identity, are prejudiced and non-prejudiced as well. In Jardina's perspective, racism and White supremacy are not intrinsic to White identity. The new Whiteness and expressed solidarity are circumstantial and cut across all social classes, not just rural working class Whites (Hochschild 2016).

There are clearly two perspectives on White identity and racism. One can be inferred from decades of definitional writing that racism is intrinsic to Whiteness; in the second, the two can be separated. In recent work on White identity politics, White identity and racism are distinct entities. In the end, we argue that White Americans have agency in choosing which *type* of White identity they assume. They can separate Whiteness from racism, and, hopefully, embrace an anti-racist White identity.

The Age of Trump

Donald Trump did not cause post-2016 racism. Since 1950, good-paying union manufacturing and assembly jobs were moved from the urban Northeast and Midwest into the American South and then overseas and to Mexico. Businesses sought lower labor and production costs in the emerging global economy. The same deindustrialization that devastated Black working class urban communities in the 1960s caught up with the rural and small-town White working and middle classes. Consequently, by the 1990s, it became apparent that many children of White Americans, who benefitted economically from post-World War II industrialization and became the backbone of the post-1950s working and middle classes, found themselves downwardly mobile, as illustrated for the State of Kansas (Frank 2004). More efficient and profitable global free markets spelled the end of small businesses, small farmers, downtown business districts, personalized multigenerational community life, and local economies.

Downward mobility is never pretty. There is a long history in the United States of contempt and utter disdain for those who fall in social class. The lowest class of Whites has been viewed by those higher up at various times in American history as filth, rubbish, crackers, poor White trash, clay-eaters, bastards, a defective breed, country, hill-billies, and rednecks (Isenberg 2016). Since 1990, each year, thousands of White Americans, who once had prospects of achieving and enjoying the American dream, dropped into the class of Americas long time "deplorables."

Their decline was made worse by almost no one noticing. Journalists, social scientists, politicians, people in social media, entertainment, and even pop culture focused instead on celebrating the affluence and lifestyles of America's emerging professional and global middle class (Brooks 2000). The excitement was over the emergence of whole new industries, i.e., biotechnology, software engineering, computer applications, artificial intelligence, medical technologies, and solar and wind power. Meanwhile, older industries that were labor-intensive declined, collapsed, or went overseas with little notice, i.e., coal mining and fossil fuel power generation, steel manufacturing, auto assembly, the garment industries, and consumer goods manufacturing. Everything made in China, Korea, Brazil, and Pakistan represents jobs lost since 1960 in urban America. By 1990, the lost jobs were now those of the White ex-urban working class.

A general social movement emerged from declassed Whites. Initial mobilization in Kansas ended abortion rights which led the movement to take control of the state's Republican Party. The movement rapidly spread. In contrast, the Democratic Party was blindsided entirely by the declassed White working class' mounting alienation and anger. One White worker is quoted as saying, "A working person who supports Democrats is like a chicken that supports Col. Sanders!" (Frank 2004). However, at its core, the movement in reaction to White downward mobility is more cultural than political and more about lost identity than addressing economic grievances. A belief of the movement is that politicians of any party never deliver, steal, lie, and are untrustworthy. So, the movement is defiant and oppositional toward established government and political parties. Politicians supported by the movement are not expected to govern; instead, they are sent to convey a harsh message of defiance to the establishment and are to be oppositional (i.e., Tea Party and Freedom Caucus). The movement is also reactionary in that followers wish to return to an idealized family-centered, God-fearing, pre-1960s, small-town America. It is anti-intellectual, anti-authority, pro-gun, affirms rigid individualism, petty-capitalism, White nationalism, and is openly racist. It is pro anything that would upset and disgust the professional middle class and the liberal establishment.

Such a movement is once in a lifetime fertile ground for America's rich traditions of fringe beliefs and fundamentalists. The vacuum in trust of established political parties has been apparent in declining voter turnouts since before 2016. Related alienation from established religious faiths has filled the pews of non-traditional churches with absolutist beliefs that provide a new sense of community. Conspiracy theorists, who until recently, were ignored and laughed at suddenly have serious followers. Skinheads, Neo-Nazis, the Alt-Right, the Ku Klux Klan, and White Nationalists, who have existed for decades on the fringes of American life, now have new life, new followers, new confidence, and enthusiastically parade in public.

However, the most potent exploitation of White alienation and anger is decades of effort by billionaire libertarians to shape White political beliefs and exploit White discontent (Mayer 2017). Since the 1960s, David and Charles Koch, Richard Scaife, John Olin, and a host of others have spent billions of dollars setting up conservative institutions, supporting conservative writers, building inroads into elite academic institutions and departments, shaping judiciaries, and funding increasingly radical conservative politicians and propositions. They have been the elite money behind the conservatism of the past five decades. However, their real intent has not been to advance historical conservatism. Their real objective has been to wrestle control of Federal and State governments away from centrist Republicans and Democrats to reduce all government to a size where it can no longer control or threaten their personal and business interests and accumulation of wealth. Their goal is to do away with regulations and taxes on businesses, eliminate social welfare and environmental interventions, and abolish all federal departments except defense and, perhaps, the State Department.

The media billionaire Rupert Murdock has played an even more central role in this movement. He is the founder and the money behind Fox News, which reflects his worldview (Stelter 2020). More than anyone else, he has exploited the anger of White Americans over their downward mobility by willfully propagating racism through traditional media. Sean Hannity and Tucker Carlson on Fox television, the late Rush Limbaugh on radio, and hundreds of other television and radio propagators provide daily doses of a "White racial us-versus-them." There is a direct relationship between the extent to which viewers exclusively watch Fox News and their increasing racial animus. More than anyone else, Rupert Murdock set the national stage for Donald Trump.

Donald Trump opened the door for discontented Whites to express the racial animus that has been just out of sight for decades. He also provided a window for students of race relations to see the actual mechanics of racism's propagation and influence on institutional life. Before Trump, it is as if those who studied racism had to theorize that racism was an unseen planet circling our sun. Then, suddenly, in 2016, the unseen planet appeared. Four developments make it essential to focus on this movement and period in this current edition of *Impacts*.

First, President Trump's failure to distance himself from Skinheads, Neo-Nazis, the Alt-Right, the Ku Klux Klan, and White Nationalists made it very clear that racism in American life is neither symbolic nor unconscious. Millions of White Trump supporters could finally express decades of pent-up-racial hatred, defying established norms. In Trump, they found a spokesperson who freed them from political correctness to say what they believed about Blacks, Latinos, and other races. After 2016, it became apparent that racism was no longer colorblind for a growing percentage of Americans, and race's centrality had not declined in significance.

Second, during the Trump Presidency, the existence of racism received almost daily affirmation. The hundreds of White men who marched with torches in Charlottesville articulated all the central premises of White supremacy and cultural racism. Since then, frequent marches and demonstrations by the Alt-Right, Proud Boys, QAnon Conspiracy Theorists, and other heavily armed White men are reminders that the Charlottesville march was not a one-time event. Blowback is not just from avowed racists. Cell-phone videos show the extent to which White police officers are without humane regard for Black lives because they fundamentally do not believe that Black people are their equals. Their disrespect and inhumane treatment are facilitated institutionally and defended by the criminal justice system and police unions. These are all examples of in-your-face institutional racism, which have, in turn, encouraged increased individual racism.

Third, we learned something new. Donald Trump showed us that integrity and leadership or rather the lack of them, matter (Cohen 2020). The federal government plays a critical institutional role in increasing or decreasing racism in American life. The President can use the Department of Education, State Department, the Department of Homeland Security, and the Department of Health and Human Services to visibly advance or retard the common good. Compromising the Justice Department and F.B.I. were particularly damaging (Strzok 2020). In which case, we are reminded that government can and should play a vital role in combatting racism. Agencies such as Environmental Protection, the U.S. Centers for Disease Control and Protection, and the Food and Drug Administration that do essential life-saving work can be rendered ineffective and used to retard racial progress by racist leadership.

Fourth, the White Americans who willingly and spontaneously demonstrated and risked injury and arrest to protest the murder of George Floyd did a lot more. They rejected arbitrary police murders of Blacks and showed, in large numbers, that they actively oppose racism. The Black Lives Matter and George Floyd demonstrations showed that, if all Whites benefit in some way from racism and race privilege, there are many who question the benefits and the price they pay for Whiteness. They speak to our central theses. Not all Whites wish to benefit from racism, and White privilege comes with material, moral, and psychological costs.

The Impact of Racism on White Americans

White Americans gain unearned privileges from racism relative to people of color. The phenomenon of White privilege is real and well documented. At the same time, White Americans also pay a price for racism. The costs of racism to Whites are institutional, cultural, and individual.

For example, consider the fervent support that Donald Trump received for his 2016 campaign promise to build a wall along the U.S.–Mexico border. With chants of "Build the Wall," Trump supporters fervently wanted to cut off all undocumented immigration from South and Central American countries. Part of their motivation was economic. Many working-class Whites have seen manual labor jobs offered to immigrants, both documented and undocumented. At the same time, part of their motivation is surely based on racism. There is also a strong desire to keep "Brown" people out in order to keep America "White." Notice, for instance, that there are no calls to build a wall on the Northern U.S. border to keep the predominantly White population of Canada out of the United States.

However, numerous attempts have been made since the Immigration, Reform, and Control Act of 1986 to shut the Southern U.S. border, but none of them have achieved their stated goal. Ironically, in fact, attempts to militarize or shut the border have actually *increased* the number of undocumented immigrants living in the United States. As the revolving door of seasonal immigration gets harder and more dangerous to shut, seasonal immigrants become permanent immigrants. Seasonal immigrants often come alone, but permanent immigrants often have their families to join them North of the border (Massey, Durand, Malone 2002).

Attempts to close the border have another effect as well; they drive down the price for manual labor. Undocumented immigrants working in the United States have zero labor rights. They have no legal recourse when their employer makes them do more work than agreed or pays them less than promised. Undocumented immigrants cannot even demand to be paid the legal minimum wage. As a result, employers take a legal risk in hiring undocumented workers, but they can save a ton in labor costs. As more and more employers turn to hire undocumented workers, the price for manual labor falls, and U.S. citizens get squeezed out of the manual labor market, increasing their economic woes. Blaming the undocumented immigrants for their economic loss (instead of the employers looking to save a buck), many Whites in the U.S. increase their demands to shut the border, and the cycle repeats itself (Massey, Durand, Malone 2002). In this way, White American racism against Latinx people South of the U.S. border ends up negatively impacting Whites themselves.

Calls to "shut the border" or "build the wall" are not the only way that racism negatively impacts White Americans. Whites earn less in racist economies than in anti-racist economies, pay more for housing in segregated communities, and suffer and die from insufficient healthcare. In addition, Whites pay more for government, and White elites cause other classes of Whites to sacrifice more for their foreign policy. Whites suffer psychologically from their position as the dominant race, and White women pay a cost for the stereotypes applied to BIPOC women. The mostly White audiences of conservative media know less about the world and their position in it. White university students and faculty suffer from racist incidents against BIPOC at their universities, and White teachers suffer from racism in their K-12 schools.

There is, however, a better option. We all have agency—can make choices and act on our social environments. We have the power to challenge elites and their interest groups that propagate racism to confuse and separate us. White Americans have the power, opportunity, and choice to accept, reject, ignore, or work to change their racial identity and racial culture. We have the same agency regarding racism. The racial social structure will continue with or without our beliefs. Undoing racism means we must dismantle the racial hierarchy.

Organization of the Book

Part I: The Costs of Racism to White Americans

Part I provides illustrations of the costs of racism on White Americans during this era of Trump. The costs are outcomes of the influence of elites on institutions. Michael Reich, in Chapter 2, updates the contemporary economics underlying White racial losses and gains from the first edition. The assumptions that all White Americans gain economically from racism are shattered along with the idea that Whites do not have substantial economic losses due to racism. In Chapter 3, Jacob Rugh exposes the increased cost of housing for White Americans due to racial segregation and White flight. In Chapter 4, Jonathan M. Metzl explores racism's costs to working- and lower-class Whites in unnecessary poor health and deaths.

In Chapter 5, Robert Fantina reviews the Trump administration's overt racism and the damage racism did in four short and tumultuous years to government agencies. In Chapter 6, Chris Danielson shows that domestic racism has had a corrosive and costly effect on U.S. foreign policy. Here again, mostly White Americans pay a high cost in for conscious and unconscious American racism overseas. Finally, racism and sexism intersect in the functioning of all institutions. In Chapter 7, Charlotte Dunham's traces the effects of gender scripts on White women's racial attitudes. Dunham shows how racism mainly damages White women's attitudes toward one another, toward women of color, and compromises their efforts to bring about gender equality.

In Chapter 8, Lisa B. B. Spanierman and D. Anthony Clark advances the social psychologies of the first and second editions of *Impacts* and provide insights on how and why White Americans might have different orientations toward Whiteness. The orientations come with implicit psychological benefits and costs. In Chapter 9, Kalemba Kizito provides a case study of conservative and other media. Fox News is in a class of its own with explicit racist narratives. We see how Trump's "patriots" were primed for costly violence. Scott Brooks, Stacey Flores, and Jorge Ballesteros did an ethnographic case study, in Chapter 10, of the outcome of a boycott by college athletes. The failure of the boycott showed the role of elites in enforcing racial hierarchy at a large public university. The student athletes paid a price, but so did a surprising number of Whites. Finally, in Chapter 11, Patricia A. Maloney's ethnography shows how cultural racism implicitly shaped outcomes for beginning teachers after five years in the classroom. These studies show how institutional racism undercuts individual and cultural functioning in both direct and subtle ways.

Part II: The Benefits of Anti-Racism

Part II focuses on the struggle to reduce if not eliminate racism. It shows that we may have to go back two steps to come forward three, and so on. In Parts I, we show how racism is propagated. In Part II, we show how the same institutions can be used to push racism back and reduce its impacts on everyone. In Chapter 12, Pamela Oliver provides a history of the struggle to reduce if not eliminate racism prior to #Black Lives Matter. Overall, there have been more gains than losses. In Chapter 13, Fred Pincus explains what affirmative action is. He dispels the conservative truism that affirmative action is about reverse discrimination against White Americans. Intentional interventions into the racial practices and outcomes of institutions are the only effective way to correct historic and contemporary racial inequalities.

In Chapter 14, we summarize the mechanics of White racism and the impacts of racism on White Americans from the previous chapters and from the two prior editions of *Impacts*. The age of Trump has provided an unprecedented view of racism's inner workings. Authors in this edition have expanded the list of insights and impacts of racism on White Americans. Finally, in Chapter 15, we conclude with new public policy recommendations, and our assessments of where we go from here.

Discussion Questions

Q1: How does the definition of racism in this chapter differ from a popular, dictionary definition of racism? Which do you think is correct, and why?

Q2: What is *White identity*, and what is *Whiteness*? Is *Whiteness* a form of cultural racism?

Q3: How was the election of Donald Trump a product of Whiteness?

Q4: What are some examples of the costs that Whites pay for being the dominant racial group?

References

Blumer, H. 1958. "Race Prejudice as a Sense of Group Position." *Pacific Sociological Review* 1 (1): 3–7.

Bonilla-Silva, Eduardo. 2018. *Racism Without Racists: Color-Blind Racism and the Persistence of Racial Inequality in America*.

Brooks, David. 2000. *Bobos in Paradise: The New Upper Class and How They Got There*. New York: Simon and Schuster.

Cohen, Michael. 2020. *Disloyal, A Memoir: The True Story of the Former Personal Attorney to the President of the United States*. New York: Skyhorse Publishing.

Dictionary, Merriam-Webster. 2021. "Racism." Merriam-Webster Dictionary. Accessed February 9. https://www.merriam-webster.com/dictionary/racism.

Frank, Thomas. 2004. *What's the Matter with Kansas? How Conservatives Won the Heart of America*. New York: Henry Holt and Company.

Hardiman, Rita. 1982. "White Identity Development: A Process Oriented Model for Describing the Racial Consciousness of White Americans." University of Massachusetts Amherst.

Hochschild, A.R. 2016. *Strangers in Their Own Land: Anger and Mourning on the American Right*. New York: New Press.

Isenberg, Nancy. 2016. *White Trash: The 400-Year Untold History of Class in America*. New York: Viking.

Jacobson, Matthew F. 1999. *Whiteness of a Different Color*. Cambridge: Harvard University Press.

Jardina, A. 2019. *White Identity Politics*. Cambridge University Press.

Kendi, I. 2019. *How to Be an Antiracist*. London: One World.

Lopez, Alfred J. 2005. *Postcolonial Whiteness: A Critical Reader on Race and Empire*. Albany: State University of New York Press.

Massey, Douglas S., Jorge Durand, and Nolan J. Malone. 2002. *Beyond Smoke and Mirrors: Mexican Immigration in an Era of Economic Integration*. New York: Russell Sage Foundation.

Mayer, Jane. 2017. *Dark Money: The Hidden History of the Billionnaires Behind the Rise of the Radical Right*. New York: Anchor Books.

McIntosh, Peggy. 1988. *White Privilege and Male Privilege: A Personal Account of Coming to see Correspondences Through Work in Women's Studies*. Wellseley College Center for Research on Women.

Merton, Robert. K. 1949. "Discrimination and the American Creed." In *Discrimination and National Welfare*, edited by Jewish Theological Seminary of America, 99–126. Port Washington, NY: Kennikat Press.

Morrison, T. 1992. *Playing in the Dark: Whiteness and the Literary*. Cambridge: Harvard University Press.

Omi, Michael, and Howard Winant. 2014. *Racial Formation in the United States*. New York: Routledge.

Roediger, David R. 1991. *The Wages of Whiteness: Race and the Making of the American Working Class*. London; New York: Verso.

Stelter, Brian. 2020. *Hoax: Donald Trump, Fox News, and The Dangerous Distortion of Truth*. New York: Simon ans Schuster.

Strzok, Peter. 2020. *Compromised: Counterintelligence and the Threat of Donald J. Trump*. Boston: Houghton Mifflin Harcourt.

Thandeka. 2020. *Learning To Be White: Money, Race and God in America*. Continuum.

Part I

The Costs of Racism to White Americans

2

Economy: Racism's Continuing Economic Cost to Whites—A Second Look

Michael Reich

Abstract The economy is perhaps the most decisive institution in a society. The ability of government to address social problems, our standard of living, and general well-being is dependent on the health of the economy. In the first edition of *Impacts*, the economist Michael Reich showed that reductions in racial economic inequality are correlated with higher earnings for Whites (and Blacks) and unionization. Reducing economic inequality is also correlated with falling industrial profits. He reaffirms these correlations in this edition. Where racial inequality increases, low- and middle-income Whites lose financially. Simultaneously, the earnings of the upper 1% of Whites increase. We learn that there are strong present-day economic incentives for Whites in the upper 1% of incomes to favor racial hierarchy. But there are also strong economic reasons for Whites in the lower and middle classes to do away with racism in the economy and in politics. Institutional racism harms most White people.

Keywords Inequality · Economy · Unions · Racism · Profits · Mobility

In 1981, my chapter "The Economic Impact in the Postwar Period," appeared in the first edition of *Impacts of Racism on White Americans*

M. Reich (✉)
Center on Wage and Employment Dynamics at the Institute for Research on Labor and Employment (IRLE), University of California, Berkeley, CA, USA

D. W. Austin and B. P. Bowser (eds.), *Impacts of Racism on White Americans In the Age of Trump*, https://doi.org/10.1007/978-3-030-75232-3_2

(1981a).[1] There I discussed the dynamic between race relations and labor relations in the post-World War II U.S. economy and the economic effects of racism on White workers. I presented statistical evidence showing that racially integrated unions had improved the lives of Black and White workers—and that racial inequality hurt White workers while benefiting White employers. Here I reprise that argument through a more historical analysis and bring it up to date.

I first sketch important positive changes that emerged in the 1930s in the relation between race relations and labor unions. In the Depression decade of the 1930s, a multi-racial coalition of Black and White workers created a new type of industrial union, resulting in economic advances for Black and White workers alike. In the postwar years, collective bargaining over wages and working conditions between employers and these industrial unions became a settled institution that structured U.S. labor markets. Throughout this period, hundreds of thousands of Black workers left agriculture to work in the industrial economy, with substantial improvements in their living standards. Wages (corrected for inflation) rose steadily for all demographic groups as the U.S. economy experienced strong levels of economic growth. And the New Deal political coalition of unions, racial minorities, and liberal professionals continued to hold political sway in the Democratic Party.

I then discuss the deterioration of race relations that begin in the late 1970s, which was only beginning to become visible when I wrote that chapter in 1981. In the economic realm, employers turned their backs on collective bargaining institutions and relocated many of their plants to the nonunion South, where racial economic inequality was also greater. In the political realm, employers became more aggressive in lobbying for lower taxes and weaker regulations. To advance their interests, Republican Party operatives and leaders disseminated racist ideologies that weakened the Democratic Party's hold on working-class Whites. Meanwhile, technological change and globalization reduced the demand for routine industrial jobs, especially relatively well-paid union jobs. These changes set the table for the period that began with Reagan and continued through the Trump years, with the further ascendance of racism, slower economic growth for most Black and White workers, and the growth of inequality. I then conclude with some thoughts about how equality-generating New Deal programs and the New Deal political coalition might each be advanced today. Here I discuss a set of economic policy reforms that would generate economic growth for most workers, while also creating greater racial inclusivity.

[1] The chapter was based on my book, *Racial Inequality: A Political-Economic Analysis* (Reich 1981b).

From the 1940s to the 1980s

From the 1940s through the 1970s, hundreds of thousands of Black and White workers left agricultural work in the rural South and entered the growing industrial economy. The integration into industry brought Black workers into greater contact with White workers and with the U.S. labor movement. The result was a new dynamic that benefited both Black and White workers.[2]

The previous relation between Black workers and unions had been problematic for Black workers and for industrial unions. Despite some scattered examples of inter-racial coalitions advocating for the interests of all workers, in the decades before the 1930s, U.S. union leaders tended to view Black workers primarily as potential competitors who would undercut White workers' jobs and wages. They also viewed Black workers as potential strike-breakers during industrial actions, as actually occurred during the steel strike of 1919. Skilled craft worker unions, such as among carpenters, plumbers, pipefitters, and electricians, also excluded Black workers from their membership, thereby reducing their access to working in those trades. This exclusionary practice was as prevalent in the North as in the South.

The dynamic between race and labor changed fundamentally in the 1930s, when new industrial unions began to organize workers in steel, automobile, electrical, rubber, glass, and other manufacturing sectors. These new unions included skilled and unskilled workers, with Black workers and White workers in racially integrated local chapters. Industrial unions bargained for the same dollar wage increases for skilled and unskilled workers, thereby narrowing the difference in pay between skilled and unskilled. Black and White workers thus experienced gains by recognizing their common economic interests. The strength of these unions was greater—because they now included all the workers in their industry, including Black workers—in integrated locals. Indeed, one of their organizing slogans was "Black and White, unite and fight."

In the decades during and after World War II, the industrial unions become an accepted part of industrial life, recognized by major industrial employers. Collective bargaining between these unions and employers became accepted, with workers getting annual wage increases that were paid for by annual productivity gains. However, these unions were unable to extend their victories into the U.S. South, where a majority of Black workers were still located. As Black workers continued to migrate to industrial cities

[2] For a detailed account of developments in this period, see Gordon, Edwards and Reich, *Segmented Work, Divided Workers* (1982).

in response to the growing demand for industrial labor, they became an increasing proportion of labor union members. In the 1960s and 1970s, Black workers in these industries were still more concentrated in unskilled jobs than White workers; and they held few leadership positions in their unions. Nonetheless, the integration of Black workers in industrial unions marked a significant advance in U.S. race relations.

A different relationship still characterized the older unions, which continued to be organized along narrow skilled craft lines; craft unions were most important in the construction industry. Workers in craft unions, who were almost entirely White, benefited by limiting entry into their occupations. Black workers made few inroads into these skilled trades. When they did, they typically had to create separate union locals for Black workers.

In this period, racial inequality thus varied substantially among individual industries. My 1981 chapter illustrated how these variations in racial inequality affected Black and White workers' pay. I showed that high levels of racial inequality generated *lower* earnings among both Black and White workers. Higher levels of racial inequality also generated higher industrial profit rates. Similarly, where racial inequality was greater, low- and middle-income White workers had lower incomes. In contrast, the incomes of the top 1% of Whites increased—and their gains exceeded the White majority's losses.

In other words, in industries with lower percentages of the workforce that were unionized, White wages were lower and business profit margins were higher. I also found that Whites in the upper 1% of incomes had more income when racial inequality was greater.

As I have already suggested, these patterns of racial inequality and changes in White incomes did not operate in a vacuum. The dynamics came largely from the new unionism of the 1930s and 1940s and particularly from the fact that these unions were racially integrated.

To summarize this point, the unions of skilled craft workers, especially in construction, mainly excluded Black workers from membership. Membership in these unions was often the ticket to jobs in these skilled craft occupations. The racial limitations on entry to the union meant that the White members had less competition when gaining skill and job opportunities. On the other hand, the large industrial unions—such as among automobile and steel workers that arose in the 1930s—included both skilled and unskilled workers in the same union locals and included Black workers.

The dynamics between race, labor, and the economy began to change in the 1960s and 1970s. The civil rights movement and the large number of Black workers employed in industry challenged unions, even skilled crafts unions, to desegregate their locals and to take Black workers' interests more seriously. At the same time, the economic booms of these decades gave unions more bargaining leverage with employers. The postwar economic and political power of industrial unions continued to grow, but then peaked in the late 1970s. In the 1980s and since, these unions were decimated. As a result, despite some exceptions—especially in some service industries—racially integrated unions no longer exert the power they had held before. What happened?

1980s to 2020

Two connected developments drastically reduced the economic and politic power of unions while also increasing racial economic inequality.[3] The first was economic in motivation. Large employers began to move factories and worksites to the nonunion South and rural areas, in what General Motors dubbed a "Southern strategy." Pay in these new locations was considerably lower than in the unionized Midwest. Business decision-makers thus pulled facilities out of cities and states with high concentrations of relatively well-paid workers—that is, with high concentrations of industrial, multi-racial unions, especially in the midwestern states. Both unionization and racial integration were much lower in the South and rural America. Relatedly, large companies in many industries no longer accepted that they had to work with unions and switched to working against them.

In the Reagan years, the weakening of unions became a deliberate part of government strategy. Already, in the latter 1970s, increases in inflation, caused in part by oil price shocks and by union-negotiated wage increases and cost of living escalators, led the Federal Reserve Board to increase interest rates sharply, to double digits. As one Reagan official in the U.S. Department of Labor put it, these moves were mean to "zap labor." These efforts then generated double-digit unemployment in the early 1980s, which in turn especially reduced the number of employed union members and further weakened union negotiating strength. Black workers in union-friendly cities were among the first to feel the effects of union and non-union job flight and rising interest rates. Globalization and technological change, which reduced demand for routine manufacturing jobs the most, further reduced the number of union members.

[3] For a more detailed account of these developments, see Reich, "Postwar Racial Income Differences: Trends and Theories" (1988).

The second development that changed the dynamic between race and the economy was explicitly political in motivation. The Democratic Party splintered in the 1970s and became less focused on the concerns of less well-off White workers. In 1980, Lee Atwater of South Carolina, then a strategist for Republican congressional campaigns in the South, instituted a political version of the "Southern strategy." His goal was to drive lower- and middle-class White working people out of the Democratic Party and into the Republican Party. By 1984, Atwater had become Ronald Reagan's deputy campaign director. Atwater succeeded in painting the Democratic Party as the political party of Black people. Although White Southerners had been leaving the Democratic Party since the 1940s, Atwater and Reagan's strategy accelerated that process enormously.

Atwater and Reagan made inflammatory racial accusations about the Democratic Party to White voters, without explicitly mentioning Black people. The messages were: If you vote for this Democrat or that one, they will let rapists and criminals out of jail, or your taxes will go to welfare recipients who don't want to work, Black welfare queens, and so on. The consistent themes were that White people would be hurt by Democrats and taken advantage of, and that Blacks would rule over Whites. This strategy of "dog-whistle" racism, or using code words to make racist statements, became a principal and very effective strategy of the Republican Party (Lopez 2014, 2019).

In the same period, well-paying jobs increasingly left the Northeast, the Midwest, and the Pacific Coast, to relocate to the South and rural America and abroad. As a result, unionization rates declined precipitously in the 1980s, especially among private-sector workers. Racial economic inequality increased and so did business profits. Ironically, many White lower- and middle-class working people found themselves voting for leaders who were against unionization, the very institution that had kept their wages high in the prior generation. They voted for leaders who favored the economic interests that kept their wages from increasing and moved their jobs to other parts of the United States or abroad. Where unions had advanced the incomes and interests of White workers, the not-so-subtle use of racial prejudice and stereotypes obscured their interests, resulting in reduced incomes. Donald Trump drew essentially upon the same strategy to win the presidency in 2016.

The Year 2021

I have described the changes in the dynamics among race, labor, and the economy in the four decades since 1981. To what extent do these relationships still hold? As I have suggested, the divide and conquer dynamics that I described in my 1981 chapter remain very much in force, perhaps more so. A series of pro-business Republican Presidential administrations (Reagan, Bush 2, Trump) have progressively reduced business taxes and weakened business regulations. We now know what the U.S. economy looks like without unions and with very weak government taxation and regulation: it is much more unequal and has produced little growth in real wages for most non-college educated workers. And the Black–White wage gap has grown.

Since 1981, the relocation of jobs from union-friendly states and cities into the country's South and rural areas continued, and as jobs also moved to Mexico, China, Vietnam, Bangladesh, and other low-wage countries. The small towns and cities that experienced plant and manufacturing closures have left many White workers without work and without any immediate prospects of finding jobs that pay a livable wage.

To add insult to injury, during the 2008 Great Recession the federal government bailed out big banks to keep the economy from crashing, but it offered little support to anyone else.[4] Homeowners who received subprime loans from these very same banks were not bailed out; Americans who were sold home loans ended up owing more than their homes were worth. Other homeowners were not able to keep up with payments because of rising interest rates. As a result, large numbers of Americans of all races lost their homes and had their credit ratings ruined. Their loss of faith in government and their anger at government remains evident today.

To summarize, since the 1980s, Black workers and the White lower and middle classes have experienced stagnating wages and reduced home-based wealth. These experiences form the economic backdrop to White Republican anger and today's political polarization in the United States. Ironically, the blame has not fallen on employers who decided to move their jobs or bankers who sold them sub-prime mortgages. It is business as usual for the top income groups who benefit from profit maximization and sub-prime lending. The Republican Party, which many White workers support, has shifted the political blame to the Democratic Party and to the Black voters who make up a large part of its base. Fox News now provides them with daily doses of dog-whistle racism—the "Southern strategy" in operation. As a result, the anger

[4] For more detail, see McDonough, Reich, and Kotz, *Contemporary Capitalism and Its Crises* (2010).

and alienation of the White working class has been separated from the sources of its economic losses. White identity politics has become more important than economic interest.

Solutions

What needs to happen to reduce racial economic inequality and make the economy work better again for all workers, regardless of race or ethnicity? Most important, the Democratic Party must become once again the clear champion of all less well-off workers, whether they are Latino, Black, White, or Asian. An economic policy that prioritizes full employment for all these groups would help enormously. The Biden Administration has indeed already proposed such policies.

And here are three labor market reforms that Congress could legislate. The first policy would increase the federal minimum wage, which has not changed since 2009 despite a 24% increase in the cost of living. Higher minimum wages benefit both Black and White workers; they especially aid Black workers to overcome the housing discrimination that has kept many from being able to access better-paying jobs (Wursten and Reich 2021). A second policy would increase unemployment insurance benefits and the length of time they are available. Currently, in the low-wages states these benefits are much lower and more difficult to get—for both Black and White workers. The federal government should take over the administration of the unemployment system and establish national benefits levels comparable to those in the highest-wage states. Such a reform would especially help Black workers, who are a larger proportion of the workers in the low-wage states, while also helping White workers. Third, reform labor laws to make it much easier for workers to form unions. The current laws make it almost impossible to organize workers into unions. As I explained earlier, stronger and racially integrated labor unions are the most effective way to improve workers' wages and reduce racial economic disparities.

These policies would go a long way to reducing racial economic inequality while also increasing broadly-based economic growth. But they would not be sufficient. The additional needed policies lie beyond the scope of economic tools and practice. We need political response to the deliberate use of White racial identity and dog-whistle racism that lie to and manipulate White people into supporting regressive social policies. These appeals have become a political threat to our democracy and economy; they require an even more effective political response. More than ever before, ongoing multi-racial political mobilization will have to be part of that response.

Discussion Questions

Q1: How does racial inequality in the economy impact White Americans?
Q2: What happens to average wages for White Americans when businesses and unions become more racially integrated?
Q3: How can unions hinder racial equity, and how can unions support racial equity?
Q4: What happens to wages for White Americans when unions support racial equity?

References

Gordon, David, Richard Edwards, and Michael Reich. 1982. *Segmented Work, Divided Workers: Historical Transformations of Labor in the U.S.* New York: Cambridge University Press.

Lopez, Ian Haney. 2014. *Dog Whistle Politics: How Coded Racial Appeals Have Reinvented Racism and Wrecked the Middle Class.* New York: Oxford University Press.

Lopez, Ian Haney. 2019. *Merge Left: Fusing Race and Class, Winning Elections, and Saving America.* New York: The New Press.

McDonough, Terrence, Michael Reich, and David Kotz. 2010. *Contemporary Capitalism and Its Crises.* Cambridge: Cambridge University Press.

Reich, Michael. 1981a. "The Economic Impact in the Postwar Period." In *Impacts of Racism on White Americans*, edited by Benjamin P. Bowser and Raymond G. Hunt, 165–76. Beverly Hills: Sage.

Reich, Michael. 1981b. *Racial Inequality: A Political-Economic Analysis.* Princeton: Princeton University Press.

Reich, Michael. 1988. "Postwar Racial Income Differences: Trends and Theories." In *Three Worlds of Labor Economics*, edited by Garth Mangum and Peter Philips, pp. 144–67. Armonk, NY: M. E. Sharpe.

Wursten, Jesse, and Michael Reich. 2021. "Racial Inequality and Minimum Wages in Frictional Labor Markets." IRLE Working Paper no. 101-21. UC Berkeley Institute for Research on Labor and Employment.

3

Housing: From Segregation to Isolation—White Americans in the Age of Trump

Jacob S. Rugh

Abstract How we organize community and housing tells a great deal about the interactions and exchanges between the people who live in them. Social trends are visible in housing before they are elsewhere. So, this is a good institution to study the effects of racism. According to Rugh, the 240 "low segregation-high education" counties in the United States represent the nation's most racially integrated and well-educated Americans. On average, they are majority-White, and closely reflect the nation's racial composition. These counties rejected Trump in 2016 and more so in 2020. Other counties that are quickly growing economically and becoming better educated and more racially integrated also rejected Trump. Rugh suggests that another trend is emerging as well. There is new evidence that home values appreciate faster in stably integrated and diverse neighborhoods than in overwhelmingly White communities. Integrated spaces are now valued more highly than all-White spaces. These trends have major implications for how White Americans will respond to the nation's changing demographics and to the continuing social cost of maintaining racial segregation.

Keywords Community · Housing · Segregation · Racism · Whites · Trump · Racism · Inequality

J. S. Rugh (✉)
Brigham Young University, Provo, UT, USA
e-mail: jacob_rugh@byu.edu

D. W. Austin and B. P. Bowser (eds.), *Impacts of Racism on White Americans In the Age of Trump*, https://doi.org/10.1007/978-3-030-75232-3_3

In the summer of 2020, after the murder of George Floyd by police officers in Minneapolis, Minnesota, over 23 million Americans participated in the largest and most racially integrated protests for racial justice, chanting the Civil Rights slogan of our times, "Black Lives Matter." Newfound White support for Black Lives Matter was mirrored by a meteoric rise in public opinion, increasing more in two weeks than it had in two years. Then, multiracial democracy in the streets was echoed by the greatest voter turnout in a century to elect Joe Biden and Kamala Harris, the first woman, Black woman, and Asian American woman as vice president as successors to Donald Trump and Mike Pence. Multiracial coalitions led by organizers like Stacey Abrams flipped the rapidly growing—and rapidly integrating—states of Arizona and Georgia and the traditional battleground states of Michigan, Pennsylvania, and Wisconsin.

Yet, White backlash would soon shock the nation's democracy at its physical and metaphorical beating heart, the United States Capitol Building in Washington, D.C. Shortly after the election, Donald Trump relentlessly lied, claiming that he won the election by a landslide, even after voters and multiple courts most certainly declared he did not. As Congress was literally about to declare Joe Biden the winner, thousands of nearly all-White Trump supporters staged an insurrection intent on capturing elected leaders and preventing the peaceful transition of power, the hallmark of U.S. democracy. Furiously self-entitled White rioters broke doors, windows, glass and attacked police while others replaced the American flag to fly a Trump banner; they desecrated the sacred chambers and declared Trump "emperor of democracy." Five people were dead, yet only a dozen or so were arrested that day.

How did the United States arrive at this point? How did a defining act of democracy (an election), and democracy itself, come under siege? This chapter offers one answer: the insurrection against the democratic republic resulted directly from the rising costs of White Americans' racial segregation. The conflict tearing White Americans apart traces its origins to the contradiction between freedom and slavery. The mounting toll of this contradiction in the last decade, the age of Trump, alone makes the stakes of White isolation as dire as ever.

After quelling the insurrection, a defiant Congress reconvened late into the night to complete the formal transition of power. However, perhaps not surprisingly, 138 members who represented the Whitest and most racially isolated districts voted to object to the election results in several states. Their brazen, openly White nationalist tone and what official votes will forever record in history were not a quirk of derangement or prejudice. What we witnessed was the product of centuries of slavery, segregation, as well as decades of redlining, gerrymandering, Southern strategizing, subprime lending, digital disinformation, profiteering, willful ignorance, and racial paranoia.

White Americans are neither a monolith nor are their racial and political views frozen in stone despite current polarization (Abrajano and Hajnal 2015). White radicalization born of segregated housing, gerrymandered White districts, and a White-dominated conservative media and online ecosystem poses a far greater risk than standard political polarization. The radicalization of White voters has been fostered, reinforced, and ultimately justified by segregation, isolation, and the lie of White supremacy. This awful dark side to the fracturing of White Americans fueled Trump's rise, from the racist roots of the Tea Party movement to the surprising acceptance by White voters of Trump's racist birther conspiracy. The awful truth confronts us repeatedly in the White nationalist terror and murder in Charleston, Charlottesville, Pittsburgh, El Paso, and, finally, in the Washington, D.C., insurrection.

The Mounting Costs of White Isolation

The political age of Donald Trump may be traced back to 2011. His launching pad was his racist birth conspiracy that Barack Obama, the first Black president, was not born in America. Why did so many White Americans believe him? I hypothesize that they did so primarily because of racial segregation in housing reinforced by radical right broadcast and online media that are even more isolating. They have little to no exposure to people of color as neighbors and peers to judge for themselves about what they are told.

The election of Trump in 2016 was a symptom of White American racism and an indicator of the crushing costs of racial segregation. The toll is already taken by housing segregation, subprime lending, and the Great Recession, which punctuated a decades-long slide in wages, union membership, and structural economic change, made more White people in White spaces vulnerable to radicalization. They are more easily threatened and are more fearful and angrier than in past decades. But this was merely the tip of the iceberg.

The hidden danger beneath the surface was the tremendous share of spatially isolated White voters who were inclined to vote for Trump, not because of the usual economic precarity, but from a fear of status threat tightly linked to racial isolation and segregation. The role of racial segregation in launching Trump to the presidency cannot be overstated. In a comprehensive study of over 125,000 White voters, researchers at Gallup found that zip code White racial isolation was one of the strongest predictors of Trump support

(Rothwell and Diego-Rosell 2016), far stronger than measures of economic dislocation, exposure to trade, or immigration.

Using data from the 2016 Cooperative Congressional Election Study, a large-scale dataset with over 60,000 verified voters, I calculated the regression-adjusted fraction of White voters who switched from voting for Obama in 2012 to Trump in 2016. Figure 3.1 shows that White voters who switched votes from Obama to Trump were activated by a context of metropolitan Black–White segregation and polarized along education/class lines and increasing White isolation relative to their county. In Fig. 3.1, Panel A shows the results in highly segregated contexts where the Black–White dissimilarity index is over 60, a standard used in prior research. Along the x-axis is the relative Whiteness of the zip code compared to counties where the White Obama voter lives. Along the y-axis is the fraction that switched their vote from Obama to Trump.

Panel A shows a higher trend line than panel B and consists of White voters in Highly Segregated Metro Areas with a Black–White Dissimilarity Index > 60 ($n = 6602$). The most striking results are shown in Panel A, which shows that the share who switched to Trump grew from less than 8% to 20% among White voters without a college degree. This growth happened when moving from the least isolated zip codes (50 percentage points less White) to the most isolated White zip codes (50 points Whiter). The slope of change is statistically highly significant. (The outcome could not be due to chance.) In stark contrast, metropolitan racial segregation and local White isolation play no statistically significant role in vote switching among three other types of White voters—those with college degrees and those without degrees outside highly segregated metro areas (Panel B). The line is essentially flat, indicating that local and metro context did not activate as much of a sense of threat and an attachment to Trump's racist appeals among these other White voters. Simultaneously, it is also true that college-educated White voters in highly segregated areas were slightly more likely to switch to voting to Trump than their counterparts in less segregated areas.

As is well known, the most segregated states in terms of housing patterns—Michigan, Ohio, and Wisconsin—also saw the highest rates of vote switching from Obama to Trump. These states paved the way for Trump's razor-thin winning margins and unexpected electoral college win. Critically, White support for Trump in 2016 is also statistically significant with housing distress (negative equity or underwater mortgages) in the wake of the midwestern (Raymond 2017; Zonta and Edelman 2016) and Florida housing crisis (Rugh 2019). These areas flipped from Obama to Trump and decided the election.

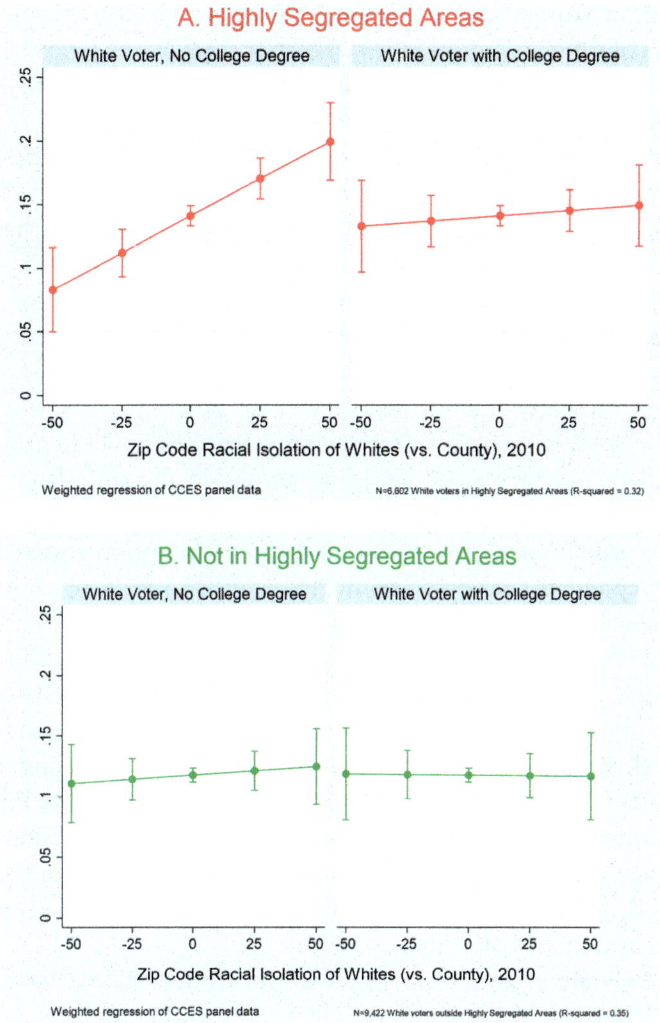

Fig. 3.1 Panel A. fraction of White voters switched support from Obama in 2012 to Trump in 2016 and Zip Code White isolation relative to county, by White voter college attainment Panel B. White Voters in Less Segregated Metro Areas: Black–White Dissimilarity Index < 60 ($n = 9422$). *Source* Calculations by the author are regression-adjusted proportions of White voters controlling for political ideology, union membership, income, and county fixed effects using the 2016 Cooperative Congressional Election Study. *Note* For rural counties, a state-level dissimilarity index score is used from the 2010 Census

In the age of Trump, further costs of residential racial segregation for White Americans continue to mount in several other areas of life. Below are findings from just five domains: social mobility, economic growth, anti-vaccination beliefs, suicide, and homicide (including police killings).

Social mobility, or the American Dream, is measured by the chance to move up from one's parent's economic level to a higher level. The likelihood of moving from a lower (parent) quintile to a higher or richest quintile (child) is significantly and substantially lower in more racially segregated metropolitan areas. The effect is not limited to people of color. Intergenerational social mobility is also lower in White zip codes and for White people raised in more segregated areas, reflecting the racial spillover in costs of segregation (Chetty et al. 2014).

Economic growth is slower in more segregated metropolitan areas, and the negative effect has increased over time alongside the shift from manufacturing to services and technology (Li et al. 2013). More racially integrated metro areas enjoy faster growth due to greater innovation fostered by more diverse entrepreneur networks and higher returns to venture capital (Samila and Sorenson 2017).

Anti-vaccination beliefs are clustered spatially in relatively affluent White communities. Their racial homogeneity leads to a false sense of protection from contagious diseases (Estep and Greenberg 2020). The tremendous toll of the novel coronavirus on White Americans in late 2020 pandemic waves reflects lower mask-wearing rates. Also, politicized and racialized disbelief in science is concentrated disproportionately in White places. White people are the oldest age group and have experienced the most deaths from COVID-19, with racial disparities waning over time as the novel coronavirus crosses state boundaries and racial color lines.

Suicide rates are at an all-time high in the United States, driven by non-Hispanic White suicides, which occur at a rate 2.5 times greater than among Black people. When we consider race and gender, suicide rates are a staggering nine times higher for White men versus non-White women (American Association of Suicidology 2019). Counties with lower racial diversity levels, social capital, and population density exhibit higher suicide rates (Recker and Moore 2016). States with the highest suicide rates are disproportionately White. Higher shares of suicides are by firearm, underscoring how higher firearm ownership and less strict gun laws in more racially isolated areas are lethal to their mostly White populations.

Fatalities involving police: While Black men today are about 2.5 times more likely to be killed by police, fatality rates are elevated for White men in areas with a legacy of racial segregation and racial violence. To be clear, Black residents bear the disproportionate burden of this legacy. Still, it is also true that Whites are more likely to be killed by police and civilians in segregated areas than in areas without that legacy. For example, Anniston, Alabama, is known for White violence against Freedom Riders in 1961 (including John Lewis), who bravely protested public transportation segregation in the South. Since 2013, the metropolitan area with the highest rate of fatalities involving police for White people is in Anniston, which is over nine times as deadly as the safest metros for White people (Schwartz and Jahn 2020). Racial disparities in police killings across the former Confederacy states are diminished because fatality rates for White people are *higher* than elsewhere, not because Black rates are lower (Schwartz and Jahn 2020). Annual homicide rates in Anniston also range between five and ten times higher than the national average. The metro areas with the highest homicide rates are also the most segregated (e.g., Memphis, Tennessee, and Baltimore, Maryland).

As the preceding data demonstrates, the costs of segregation for White people are a matter of life and death and are one source of the White isolation that fueled Trump's rise that has brought our nation to the brink.

Racial Integration, Education, and the 2020 Election

How will the White isolation of the age of Trump live on after his presidency? People of color, despite internal differences, are far less divided than White people today on matters of racial equality. The upside of Whiteness' historic fracturing is that White people joined Black people in 2020 in proportions never seen before in American history (Alexander 2020). Black Lives Matter's unprecedented resonance in 2020 and the way Black votes mattered in tipping close election outcomes Georgia and elsewhere galvanized a new multiracial coalition. This resonance begs a pair of final questions: Is there a pattern to White people joining people of color in rejecting Trump and his White nationalist agenda? Most importantly, how does it relate to housing integration, Whites and people of color living on the same block, and children attending the same schools?

During the political age of Trump that began a decade ago, White voters have had two chances to choose between Trump in the 2016 and 2020 presidential elections. We await the 2020 data on individual voters as we waited to study the White support dimensions for Trump in 2016. Meanwhile, we can glean powerful insights from county-level voting patterns. These data reveal the role of White racial integration and education in paving the way for exactly where and why White opposition to Trump grew in 2016 and 2020. Specifically, Fig. 3.2 shows the average county-level swing in the presidential vote from 2012 to 2020 weighted by total votes cast for 1385 counties with at least 1000 Black residents and 10,000 total residents. Counties are sorted into four categories based on two criteria in combination: whether Black–White segregation is high or low (above or below median score of 45) and whether college attainment of the adult population is high or low (above or below the median of 24% of adults 25 or older).

While many segregation measures exist, for this chapter, we are most interested in the structural and spatial separation between Black and White voters into separate residential neighborhoods (Rugh and Massey 2014). All county categories are on average majority White; one stands out as disproportionately White in terms of its average racial composition—the high segregation-low education counties are 71% White as of 2019, compared to 57–61% White for the three remaining categories of counties.

As Fig. 3.2 demonstrates, there is an incredible polarization of counties by their combined segregation-education status. Among 417 "high segregation-low education" counties, in predominantly rural and exurban Midwest and Appalachian counties, voters swung 10.4 points more Republican versus the 2012 election baseline. For 488 "low segregation-low education" counties,

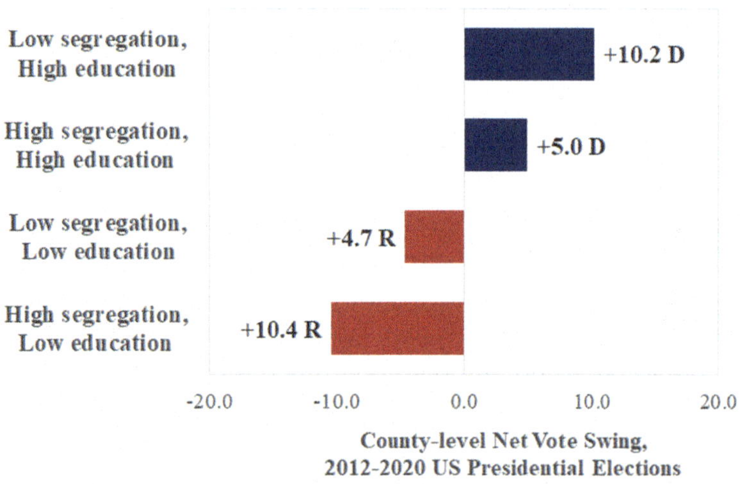

Fig. 3.2 Net Swing in Presidential Election Vote, 2012–2020 by County Category. N = 1385 counties with at least 1000 Black residents and total population greater than 10,000. Low/high Black-White neighborhood housing segregation is defined as dissimilarity index score below/above median (0.45) in 2018. Low/high education is college attainment for adults age 25 or older below/above median (24%) in 2018

predominant in the Southeast, the average presidential vote has swung 4.7 points more Republican during Trump's age. There are 340 "high segregation-high education" counties, predominant in blue states of the Northeast and suburban Michigan and Wisconsin; these counties swung 5.0 points more Democratic, away from Trump (Fig. 3.2).

Most encouraging, 240 "low segregation-high education" counties represented the most racially integrated and well-educated parts of the nation. Again, these counties are on average majority-White, the closest reflection of the nation's racial composition. They have increasingly rejected Trump both times he has been on the ballot, swinging 10.2 points more Democratic since 2012. The tremendous shift away from Trump in 2016 and 2020 among White voters in more racially integrated and highly educated areas was most pronounced in the suburban rings of large metro areas in swing states—Atlanta, Georgia; Austin, Dallas, and Houston, Texas; Denver, Colorado; Phoenix, Arizona; Minneapolis, Minnesota; Raleigh, North Carolina; Richmond, Virginia; and especially the northern Virginia suburbs of Washington, DC.

Equally notable were the large swings in the rapidly growing, highly educated, and racially integrated suburban counties that receive less attention due to their location in less electorally competitive states. The areas with the most remarkable shifts away from Trump in 2016 and 2020 include the suburbs of Greenville, South Carolina; Indianapolis, Indiana; Kansas City, Missouri; Nashville, Tennessee; Portland, Oregon; Sacramento, California; Salt Lake City, Utah; DuPage County, Illinois; and Howard and Montgomery counties in Maryland.

Indeed, the great polarization under the age of Trump has a dark side in rising support for White nationalism that is strongly concentrated geographically in less educated, highly segregated counties that are disproportionately White. This trend presents strong evidence that between two elections, Trump's support online, in conservative media, and in Congress' halls is predicated on the social fact of residential isolation of White people from Black people in America. On the other end of the scale, the highly educated, less segregated counties are nearly all suburban and rapidly growing and thriving in the post-Great Recession era. This configuration may reduce the sense of racial threat posed by greater proximity to people of color, no doubt enhanced by the fact that Black–White integration is facilitated and made more stable by Asian American and Latino neighbors (Bader and Warkentien 2016; Logan and Stults 2011). This promising trend has taken hold in nearly every state and is most apparent in suburban Arizona, Virginia, Georgia, and

Texas. These fast-growing Sunbelt states are not just rapidly integrating but also gaining electoral college votes and power.

Extreme White isolation is no longer supremely valued in local real estate; a historic shift accompanies more widespread racial integration, rising education, and higher incomes despite persistent segregation. After a century of Whiteness as property, more nuance has emerged amidst growing diversity. Local and national studies report that home values appreciate faster in stably integrated and diverse neighborhoods than in overwhelmingly White communities (Immergluck et al. 2019; Moye and Thomas 2018). But Black spaces continue to be devalued unless and until gentrification leads to broader White appeal (Hyra and Rugh 2016; Immergluck et al. 2019). Moreover, while durably integrated spaces are valued more highly than White spaces, White residents (often the numerically dominant group) benefit the most, even if the benefits are more equal than in more segregated spaces (Howell and Korver-Glenn 2020). The upshot is that White isolation is accruing a more direct and tangible cost, in stark contrast to racially integrated places. Those who live in diverse suburbs displaced White isolation of twentieth-century suburbia founded on redlining and racial discrimination. They have provided a path forward for rejecting Trump and, hopefully, White ignorance, White nationalism, and White supremacy. Only time will tell.

Conclusion: The Ultimate Costs of White Isolation

We live in a divided nation still centered on Whiteness because White Americans remain the most divided over racial justice and equality today, not people of color. After rising to 60% in June, White support for Black Lives Matter fell to 45% by September 2020, a decline nearly entirely driven by conservative White Americans, who live in more racially isolated places (Thomas and Menasce Horowitz 2020). Black and Native Americans have overwhelmingly and consistently been supportive of Black Lives Matter and progressive Democratic candidates and platforms. Nearly 9 in 10 Black Americans continue their support for Black Lives Matter, the Civil Rights movement of our time, even as its antagonists distorted, politicized, and weaponized it to erode White support in 2020 (Thomas and Menasce Horowitz 2020). Electorally, about 19 in 20 Black women have supported Democratic candidates and progressive agenda during Trump's age, followed by clear majority support among Black men, Native American, Asian American, and Latina women voters. Divides in the Latino vote have returned to a historical norm and are driven primarily by greater proximity to Whiteness regarding immigration status, nationality, class, or White male privilege.

Ultimately, White people—and Whiteness—have fractured in the age of Trump. A segment of White people has come to terms with their denial of truth, and the conflict echoes and revives a debate about the contradictions of White America when it comes to racism and freedom. Another segment of White people digs deeper into the myth of White supremacy. White fracturing and division have imposed lasting harm on American institutions and democracy. And that harm stems largely from how White people have engineered their racial isolation and growing radicalization, which now threats democracy.

Perhaps the highest cost to White people of racial segregation in housing—and, by extension, schools, workplaces, and places of worship—is a stunning poverty of knowledge surrounding the truth of our nation's racial past that perpetuates White supremacy as an imminent threat to multiracial democracy (Serwer 2021). Willful White ignorance of systemic racism, past and present, breeds denial and erodes White people's agency and freedom, and perpetuates racial inequality. James Baldwin remarked in his essay, *The White Problem*, "I am not talking about the crime: I am talking about denying one what does. This is a much more sinister matter" (Glaude 2020: 39). Baldwin was right. Far too many White people still deny the current White racism in the age of Trump, and this denial makes the problem worse.

In her path-breaking book, *White Kids: Growing up with Privilege in a Racially Divided America*, Margaret Hagerman (2018) finds that most White children accept and perpetuate the belief that White people and, notably, White spaces, are superior. However, she also uncovers a vocal minority of White children who reject colorblind racism and reveal a powerful clue to the path forward. These White children all attended integrated schools. This context helped empower them to choose to reject colorblind racist explanations of their parents, halting the intergenerational cycle of White ignorance and racism.

In other words, these White children made the antiracist choice themselves. Still, the context of racial integration and a complete education with a diverse set of peers and voices was nothing less than essential. A far taller order is to change fixed adulthood mindsets. The surge in support for Black Lives Matter and the resounding rejection of Trump over the past five years among more educated Whites in more racially integrated places provides hope that White adults can start to overcome a legacy of White superiority and White ignorance.

However, changes in White attitudes alone will never erase the towering racial deficits in income, education, and wealth. Such changes demand criminal justice and immigration reforms, among others. Indeed, the contextual arrangement of racial integration and greater higher education levels—a product of intentional desegregation and affordable housing policies in post-Civil Rights America—is a potent combination on the rise but still woefully far too rare for White children. History tells us that structural change leads to shifts in White racial attitudes and the unprecedented multiracial alliance seen in the racial justice protests of 2020. As Ibram Kendi (2017) argues, only durable, antiracist, institutional policy change will lead to racial equality and a subsequent shift in racial attitudes.

After enduring decades of opposition from White moderates, Dr. Martin Luther King, Jr., wrote in 1967, "Whites, it must frankly be said, are not putting in a similar mass effort to reeducate themselves out of their racial ignorance. It is an aspect of their sense of superiority that the White people of America believe they have so little to learn." White people must realize that they are the only ones who must take further steps onto an even longer road. They must recognize our racial past and make lasting policy changes. These include forms of reparations, truth, and reconciliation. Multiracial democracy will never flourish until more White Americans realize that it needs to become a way of life, from how they live to how they are educated. If Trump's age shows us anything, it is that White isolation, White ignorance, and the myth of White supremacy pose an existential threat to democracy itself.

Discussion Questions

Q1: What happens to home values in racially integrated areas compared to racially segregated areas?

Q2: How does intergenerational social mobility differ for White Americans in racially segregated and racially integrated areas?

Q3: What neighborhood characteristics are correlated with anti-vaccination beliefs and politicized disbelief in science?

Q4: How would police and gun reforms benefit White men in the United States?

References

Abrajano, Marisa, and Zoltan L. Hajnal. 2015. *White Backlash: Immigration, Race, and American Politics*. Princeton: Princeton University Press.

Alexander, Michelle. 2020. "America, This Is Your Chance." *The New York Times*.

American Association of Suicidology. 2019. "USA Suicide: 2018 Official Final Data."

Bader, Michael, and Siri Warkentien. 2016. "The Fragmented Evolution of Racial Integration since the Civil Rights Movement." *Sociological Science* 3: 135–66.

Chetty, Raj, Nathaniel Hendren, Patrick Kline, and Emmanuel Saez. 2014. "Where Is the Land of Opportunity? The Geography of Intergenerational Mobility in the United States." *The Quarterly Journal of Economics* 129(4): 1553–623.

Estep, Kevin, and Pierce Greenberg. 2020. "Opting Out: Individualism and Vaccine Refusal in Pockets of Socioeconomic Homogeneity." *American Sociological Review*.

Glaude Jr., Eddie S. 2020. *Begin Again: James Baldwin's America and Its Urgent Lessons for Our Own*. New York: Random House.

Hagerman, Margaret A. 2018. *White Kids: Growing up with Privilege in a Racially Divided America*. New York: New York University Press.

Howell, Junia and Elizabeth Korver-Glenn. 2020. "The Increasing Effect of Neighborhood Racial Composition on Housing Values, 1980–2015." *Social Problems*, September 4.

Hyra, Derek and Jacob S. Rugh. 2016. "The US Great Recession: Exploring Its Association with Black Neighborhood Rise, Decline and Recovery." *Urban Geography* 37(5).

Immergluck, Dan, Stephanie Earl, and Allison Powell. 2019. "Black Homebuying after the Crisis: Appreciation Patterns in Fifteen Large Metropolitan Areas." *City and Community* 18(3): 983–1002.

Kendi, Ibram X. 2017. *Stamped from the Beginning: The Definitive History of Racist Ideas in America*. New York: Random House.

Li, Huiping, Harrison Campbell, and Steven Fernandez. 2013. "Residential Segregation, Spatial Mismatch and Economic Growth across US Metropolitan Areas." *Urban Studies* 50(13): 2642–60.

Logan, John R. and Brian J. Stults. 2011. *The Persistence of Segregation in the Metropolis: New Findings from the 2010 Census*: U.S. Bureau of Census.

Moye, Richard, and Melvin Thomas. 2018. "Race and Housing Values: What Happens When Whites Don't All Move Out?" *City and Community* 17(1): 109–33.

Raymond, Elora. 2017. "The Ongoing Housing Crisis and the Rust Belt Revolt for the Gop." *Metropolitics*, January 13.

Recker, Nicholas L., and Matthew D. Moore. 2016. "Durkheim, Social Capital, and Suicide Rates across US Counties." *Health Sociology Review* 25(1): 78–91.

Rothwell, Jonathan T., and Pablo Diego-Rosell. 2016. *Explaining Nationalist Political Views: The Case of Donald Trump*. SSRN 2822059.

Rugh, Jacob S. 2019. "Vanishing Wealth, Vanishing Votes? Latino Homeownership and the 2016 Election in Florida." *Journal of Ethnic and Migration Studies* 46(18): 3829–54.

Rugh, Jacob S., and Douglas S. Massey. 2014. "Segregation in Post-Civil Rights America: Stalled Integration or End of the Segregated Century?" *Du Bois Review: Social Science Research on Race* 11(4): 205–32.

Samila, Sampsa, and Olav Sorenson. 2017. "Community and Capital in Entrepreneurship and Economic Growth." *American Sociological Review* 82(4): 770–95.

Schwartz, Gabriel L., and Jaquelyn L. Jahn. 2020. "Mapping Fatal Police Violence across US Metropolitan Areas: Overall Rates and Racial/Ethnic Inequities, 2013–2017." *PLoS ONE* 15(6): 1–16.

Serwer, Adam. 2021. "The Capitol Riot Was an Attack on Multiracial Democracy." *The Atlantic*, January 7.

Thomas, Deja, and Juliana Menasce Horowitz. 2020. "Support for Black Lives Matter Has Decreased since June but Remains Strong among Black Americans." Pew Research Center: Factank.

Zonta, Michela, and Sarah Edelman. 2016. *The Role of Midwestern Housing Instability in the 2016 Election*. Washington, DC: Center for American Progress.

4

Health: Dying of Whiteness

Jonathan M. Metzl

Abstract In this chapter, Jonathan Metzl directly connects racism with the recent phenomenon of rising White death rates, declining health indices, and increasing poverty. Metzl explains how state policies shaped by White supremacy increase mortality rates in much the same way as other manmade health risks, such as pollution.

Keywords White · Death rate · Working class · Resentment · Racism · Prevention · Health

In early 2016, I met Trevor, a forty-one-year-old uninsured Tennessean who drove a cab for twenty years until worsening pain in the upper-right part of his abdomen forced him to see a physician. Trevor learned that the pain resulted from an inflamed liver, the consequence of "years of hard partying," and the damaging effects of hepatitis C. When I met him at a low-income

Reprint from Jonathan Metzl's article with the same title published in the *Boston Review* at http://bostonreview.net/race/jonathan-m-metzl-dying-whiteness.

J. M. Metzl (✉)
Department of Medicine, Health, and Society, Vanderbilt University, Nashville, TN, USA
e-mail: jonathan.metzl@vanderbilt.edu

D. W. Austin and B. P. Bowser (eds.), *Impacts of Racism on White Americans In the Age of Trump*, https://doi.org/10.1007/978-3-030-75232-3_4

housing facility outside Nashville, Trevor appeared yellow with jaundice. He ambled with the help of an aluminum walker to alleviate the pain he felt in his stomach and legs.

Debates raged in Tennessee around the same time about the state's participation in the Affordable Care Act and the related expansion of Medicaid coverage. Had Trevor lived a thirty-nine-minute drive away in neighboring Kentucky, he might have topped the list of candidates for expensive medications called polymerase inhibitors, a life-saving liver transplant, or other forms of treatment and support. Kentucky adopted the ACA and began the expansion in 2013, while Tennessee's legislature repeatedly blocked Obama-era health care reforms.

The White body that refuses treatment rather than supporting a system that might benefit everyone is a metaphor for the nation's decline as a whole.

Even on death's doorstep, Trevor was not angry. In fact, he staunchly supported the stance promoted by his elected officials. "Ain't no way I would ever support Obamacare or sign up for it," he told me. "I would rather die." When I asked him why he felt this way even as he faced severe illness, he explained: "We don't need any more government in our lives. And in any case, no way I want my tax dollars paying for Mexicans or welfare queens."

At the most basic level, Trevor died of the toxic effects of liver damage caused by hepatitis C. Yet, Trevor's deteriorating condition also resulted from dogma's toxic effects. The dogma told him that governmental assistance in any form was evil and not to be trusted, even when the assistance came in the form of federal contracts with private health insurance or pharmaceutical companies or from expanded communal safety nets. Dogma that, as he made abundantly clear, aligned with beliefs about a racial hierarchy that overtly and implicitly aimed to keep White Americans hovering above Mexicans, welfare queens, and non-White others. Dogma suggests to Trevor that minority groups received lavish benefits from the state, even though he lived and died on a low-income budget with state assistance. Trevor voiced a literal willingness to die for his place in this hierarchy rather than participate in a system that might put him on the same plane as immigrants or racial minorities.

Trevor also died because the dogmas and hierarchies he supported reflected politicians' agendas who clamored that health care reform and Medicaid expansion represented everything from government overreach to evil incarnate. Anti-ACA invective found particular champions in GOP lawmakers in Tennessee, a once centrist state that turned hard right. These politicians repeatedly made sure that Tennessee did not create its own Obamacare exchange, expand Medicaid, or embrace the health care law in any way.

Thus, routine screenings, filled prescriptions, visits to doctors' offices, and many other factors linked to better health outcomes rose steadily in Kentucky in the four years after that state expanded Medicaid. Such trends lifted the overall well-being of many Kentuckians and particularly helped people who suffered from what are oddly called *preexisting conditions* such as hepatitis C. Oddly, in my opinion, "preexisting" assumes that a person's existence begins at the consummation of health insurance coverage. Meanwhile, preventive care and proper treatment remained unattainable for many lower-income Tennessee citizens, in large part because of their state's political choices.

As a physician, I had some sense of the complex medical and psychological explanations for Trevor's symptoms. During my internship, I worked in an intensive care unit, where I saw firsthand the devastating effects of organ failure. I then trained in psychiatry, where I realized how people's deep defense mechanisms and projected insecurities could lead them to act in ways that seem at odds with their longevity.

The more I spoke with Trevor, the more I realized how his experience of illness and his particular form of White identity resulted from his own thoughts and actions and his politics. Local and national politics claimed to make America great again—and, tacitly, White again—on the backs and organs of working-class people of all races and ethnicities, including White supporters. Politics make vague mention of governance strategies but ultimately shredded safety nets and provided massive tax cuts that benefited only the very wealthiest persons and corporations. All too often, politicians gained traction by playing to anxieties about White victimhood about imagined threats posed by "Mexicans and welfare queens."

> Ain't no way I would ever support Obamacare or sign up for it. No way I want my tax dollars paying for Mexicans or welfare queens.

Middle America

Between 2013 and 2018, I traveled extensively in the South and Midwest—in Sarah Palin's words, the "real America." I wanted to learn how people balanced anti-government and pro-gun attitudes while at the same time navigating lives impacted by poor health care, widening gun-related morbidity, and underfunded public infrastructures and institutions. During my travels, I met many people who, like Trevor, were dying in various overt or invisible ways due to political beliefs or systems linked to the defense of White "ways of life" or concerns about minorities or poor people hoarding resources. The stories these people told me became jumping-off points for a more sustained investigation of how particular U.S. notions of Whiteness—notions shaped by politics and policies and institutions, history, media, economics, and personal identities—threaten White well-being.

I found that Trump supporters were often willing to put their own lives on the line to support their political beliefs. As a result, when viewed more broadly, actions that may have seemed from the outside to be crazy, uninformed, or self-defeating served larger political aims. Had southerners, including Trevor, embraced the Affordable Care Act and come to depend on its many benefits, it would have been much harder for politicians such as Trump to block or overturn health care reform. By design, vulnerable immigrant and minority populations suffered the consequences in the most dire and urgent ways. Yet the tradeoffs made by people like Trevor frequently and materially benefitted people and corporations far higher up the socioeconomic food chain—whose agendas and capital gains depended on the invisible sacrifices of low-income Whites.

The White body that refuses treatment rather than supporting a system that might benefit everyone then becomes a metaphor for, and the parable of, the threatened decline of the larger nation. Rather than landing a man on the moon, curing polio, inventing the Internet, or promoting structures of world peace, a dominant strain of the electorate voted in politicians whose platforms of American greatness were built on embodied forms of demise.

My book *Dying of Whiteness*, which came out of this research project, explores the effects of what became central GOP policy issues. They include loosening gun laws, repealing the Affordable Care Act, or enacting massive tax cuts that primarily benefited the wealthy and corporations—on White population-level health. I also track the health effects of what various authors have called anti-government, anti-tax, pro-gun, and oft-Republican forms of

"White backlash conservatism." This is a dynamic illustrated by Trevor's rejection of the ACA because of concerns about non-White minorities taking away his resources.

My research enabled me to assemble a narrative about how, in five steps, the embrace of White identity politics by U.S. voters produced catastrophic health outcomes for those same voters.

Rather than landing a man on the moon, curing polio, inventing the Internet, or promoting structures of world peace, our era's vision of American greatness is built on embodied forms of demise.

First, a host of conservative political movements emerged (or reemerged) in southern and midwestern states over the latter twentieth and early twenty-first centuries. They brought into mainstream U.S. politics once-fringe agendas, such as the starving government of funding, dismantling social programs, or allowing the free flow of most firearms types. These movements—ranging from the Tea Party to iterations of libertarianism funded by the Koch brothers, to the Freedom Caucus, to the so-called alt-right given voice through outlets such as Breitbart—arose from vastly different agendas and points of origin. However, their interests grew ever-more aligned as they came to power in southern and midwestern states in ways that shaped state agendas, national GOP platforms, and, ultimately, the Trump administration policies. As this played out, theories of backlash conservatism gave way to something even more powerful: practices of backlash governance.

Second, these increasingly unified forms of conservatism advanced politically through overt or implicit appeals to what has been called *White racial resentment*. In other words, these agendas gained support by trumpeting connections to unspoken or overt claims that particular policies, issues, or decisions also served to defend or restore White privilege or quell threats to idealized notions of White authority represented by demographic or cultural shifts. This was both a top-down process (politicians used racial resentment as a tool for class exploitation) and a bottom-up one (the language of White resentment became an increasingly accepted way of talking about Whiteness more broadly).

To be sure, groups such as the Tea Party rose to prominence for a wide array of cultural, economic, and religious reasons, many of which had relatively little to do with Whiteness or race. Lower-income communities left behind by globalizing economies, disenchantment with Democrats, and the growing influence of corporate lobbies and megarich donors on party politics unquestionably played major roles. When I explained my thesis, some people I spoke with told me positions that appeared to reflect racism instead reflected a larger, color-blind "hatred of the poor."

Yet a major part of these movements' appeals lay in rallying cries that tapped into emotionally and historically charged notions that White Americans should remain atop other racial or ethnic groups in the U.S. social hierarchy, or that White "status" was at risk. This is not to say that anyone specific person was expressly racist. Rather, White racial resentment frameworks shaped debates about, and attitudes toward, various public policies and legislation acts. Sometimes, the racial agendas of these calls to arms were overt and obvious. For instance, then-President Barack Obama's posters photoshopped with a feather headdress, and a bone through his nose began to appear at anti-ACA Tea Party rallies. In 2016 former Missouri Republican Party director Ed Martin told a cheering Tea Party-for-Trump rally in Festus, Missouri: "Donald Trump is for Americans first. You're not racist if you don't like Mexicans." That same year, the Tea Party Patriots funded Asia-bashing advertisements featuring fictional Chinese executives in suits speaking Mandarin and laughing about how they could buy thousands of acres of Missouri farmland. At other times, the agendas' racial underpinnings appeared all but invisible to people on the ground, as with decisions to rally around issues such as guns, health insurance, or public schools—issues whose racially charged histories had been obscured by the passage of time.

Third, the policies that took shape when these once-fringe forms of conservatism entered the mainstream GOP and assumed legislative power often negatively affected the health of middle- and lower-income populations. While some of these policies and actions directly affected health care, others not expressly linked to health, such as the proliferation of civilian-owned firearms, nonetheless carried profound medical implications. White backlash politics gave specific White populations the sensation of winning, particularly by upending minorities and liberals' gains, yet the victories came at a steep cost. Increasing death rates resulted when White backlash policies became laws. There were cutting away health care programs and infrastructure spending, blocking the expansion of health care delivery systems, defunding opiate-addiction centers, spewing toxins into the air, and enabling guns in public spaces.

White America's investment in maintaining an imagined place atop a racial hierarchy harms their aggregate well-being as a demographic group, thereby making Whiteness itself a negative health indicator.

Fourth, many middle- and lower-income people experienced negative health consequences from these policy decisions. The policies involved elaborate strategies for tearing down community structures for middle- and lower-income Americans, but they hardly provided blueprints for building them back up. Minority and immigrant communities, often the targets of backlash's ire, suffered greatly and needlessly. But the health and well-being of White Americans suffered from the health effects of these policies as well. Such effects played out in public ways—such as when White concert-goers died in high-profile mass shootings linked to gun policies (or lack thereof) enacted by conservative White politicians. Other effects were far less obvious, such as the long-term implications of blocking health care reform or defunding schools and infrastructure.

Reflections

Finally, as with Trevor, many lower- and middle-income White Americans continued to support these policies and ideologies—with their inherent links to narratives of imagined victimhood and domination. They did so even after their negative effects became apparent and politicians' promises such as Trump's were unraveled. Indeed, for various reasons, White Americans in parts of the United States saw drops in life expectancy. But instead of scrapping these state-level policies as examples of historically bad governance, they became the foundations for legislation at the national level, in the form of Trump-era tax bills, gun policies, health care strategies, and other ill-fated initiatives. All the while, guns, health care, taxes, and other issues accrued larger symbolic or moral meanings in ways that rendered conversations about the effects of specific policies ever-more difficult.

The confluence of these trajectories has led to a perilous state of affairs: a host of complex anxieties has prompted increasing numbers of White Americans such as Trevor to support right-wing politicians and policies, even when these policies actually harm White Americans at growing rates. As these policy agendas spread from southern and midwestern legislatures into the halls of Congress and the White House, ever-more White Americans are then, literally, dying of Whiteness. This is because White America's investment in maintaining an imagined place atop a racial hierarchy—that is, an investment in the sense of Whiteness—ironically harms the aggregate well-being of U.S. Whites as a demographic group, thereby making Whiteness itself a negative health indicator. We make a wrong turn when we try to address racism mainly as a disorder of people's brains or attitudes. Instead, racism matters most to health when it shapes politics and policies that then affect public health.

My focus on the health effects of U.S. backlash politics for White Americans is in no way meant to minimize the larger effects of racism in the United States. It should be taken as a matter of fact, but all too often is not, that systems in which race correlates with privilege have devastating consequences for minority and immigrant populations. Cuts to health delivery networks, communal safety nets, schools, and social services, alongside policies that enable the proliferation of guns, often impact minority populations first and most severely. Racism itself can also have profoundly negative health consequences. Epidemiologist Yvette Cozier and her colleagues have uncovered associations between frequent experiences of racism—such as receiving poor service in restaurants and stores or feeling unfairly treated on the job or by the police—and higher risks of illness and obesity among African American women. Sleep researcher Michael Grandner has found links between perceived racism and sleep disturbances. And public health scholar Mario Sims found that lifetime discrimination was associated with greater rates of hypertension among adult African Americans.

Increasingly, we now hear that people with racist attitudes fare poorly as well. Racist views make people "sick" and "unhealthy," neuroscientists claim, because the psychological effort of discrimination can raise blood pressure or cortisol levels and heighten the risk for heart attacks or strokes. "Harboring prejudice may be bad for your health," neuropsychologist Elizabeth Page-Gould writes because racially prejudiced people experience such "bodily reactions even during benign social interactions with people of different races."

But my findings suggest that we make a wrong turn when we try to address racism mainly as a disorder of people's brains or attitudes or try to "fix" the problem by merely attempting to sensitize people or change their minds. On an aggregate level, people's racial attitudes have relatively little correlation to their health. Yes, in extreme cases like that of Trevor, racial animus can lead to medical disaster. Yet this correlation rarely holds true at the level of population health. Racial animosity rarely makes a person sick in and of itself—otherwise, there would be many more sick people of all backgrounds in the world.

Instead, racism matters most to health when its underlying resentments and anxieties shape larger politics and policies and affect public health. I say this in part because many of the middle- and lower-income White Americans I met in my research were not expressly or even implicitly racist. Race did not even come up in many of our conversations. Yet racism remained an issue, not because of their attitudes but because they lived in states whose elected officials passed overly permissive gun policies, rejected health care reform, undercut social safety net programs, and a host of other actions. In these and other instances, racism and racial resentment functioned at structural levels and in ways that had far broader effects than the kinds of racism that function in people's minds.

Understanding why conservative White Americans vote in ways that negatively affect their lives involves far more than pointing out ways that these voters may have been conned or deceived.

Addressing racism structurally allows me to raise what became the most troubling findings of my research: I found that, when tracked over time, racially driven policies functioned as mortal risk factors for all people living in states that had adopted them. This is because illness and death patterns that followed actions such as expanding gun proliferation or massive tax cuts mimicked those once seen in relation to other human-made pathogens, such as water pollution, secondhand smoke, or not wearing seat belts in cars, or during disease outbreaks. Society mobilized to reduce risk and improve health when toxins dumped into the water, cigarettes, or faulty automobiles led to declining health. But when the pathogens were policies and ideologies, they instead laid the foundations for politics furthered at the national level by the GOP, the NRA, and the Trump administration. In these ways, stories such as Trevor's come to embody larger problems of an electorate that, in its worst moments, votes to sink the ship they are on (except for a few privileged passengers who get lifeboats). The alternative is to invest in communal systems that might lift all boats. Anti-Blackness, in a biological sense, then produces its own anti-Whiteness. An illness of the mind weaponized onto the body of the nation.

Understanding why conservative White Americans vote in ways that negatively affect their lives involves far more than pointing out ways that these voters may have been conned or deceived. The particular issues about which Trump supporters appear to have been "duped" also tap into more extensive histories, myths, and ideologies. These histories, myths, and ideologies go a long way toward explaining the complex tension between promises of restored "greatness" on the one hand and practices of self-sabotage on the other. Better awareness of this paradoxical tension might allow us to better promote an alternative investment in collaboration and equality—in many instances, by addressing ideologies of Whiteness head-on rather than by proxy.

However, the electorate has chosen a regime whose policies come cloaked in the promise of restored privilege, enacted through polarization and divisiveness mechanisms. As a result, we talk about eliminating financial safety nets and social support programs, allowing ever-more guns, and defunding roads and bridges while at the same time enacting tariffs and building walls. Such talk, and the policies that flow from it, often signify protection, preservation, or continued supremacy. But in many instances, they ultimately serve to hemorrhage our collective abilities to solve problems or help people in times of need. Ultimately, when White voters are asked to defend Whiteness, Whiteness often fails to defend, honor, or restore them.

Discussion Questions

Q1: What reasons do poor and working-class Whites often give for opposing governmental benefits such as the Affordable Care Act?
Q2: How do White death rates, health indices, and poverty correlate with opposition to governmental benefits?
Q3: How do suicide rates for White men differ in states with strict gun laws compared to states with lenient gun laws?

Reference

Metzl, Jonathan M. 2019. *Dying of Whiteness: How the Politics of Racial Resentment Is Killing America's Heartland.* UK: Hachette.

5

Government: Calling on Racism to Run Federal and State Governments

Robert Fantina

Abstract Robert Fantina looks at the federal government's management during the Trump administration, and its effects. Americans have never known what it would be like to have segments of the federal government rendered inoperable. The Trump administration was a quasi-experiment in not having a functioning government. Trump's cabinet consisted of individuals from the top 1% of Americans who dominate the nation's wealth and are the only class that benefits financially from racism. They controlled key federal departments—education, environment, and state. These one-percenters are already overwhelmingly represented in the Senate. During Trump's administration, cutbacks in staff, budgets, and willingness to execute department missions diminished virtually all consumer, environmental, educational, financial regulations, and services. Instead, there were predictable tax breaks for the very wealthy. The cutbacks visibly impacted the poor. Less visible were the effects on working- and middle-class Americans of all races, notably the majority of White Americans.

Keywords Trump · Federal government · Power elites · White Americans · Racism

R. Fantina (✉)
University of Waterloo, Kitchener, ON, UK

D. W. Austin and B. P. Bowser (eds.), *Impacts of Racism on White Americans In the Age of Trump*, https://doi.org/10.1007/978-3-030-75232-3_5

Cultural racism has played a significant role in U.S. governance from the start and is a major driver of today's so-called wars on drugs and terrorism. It has gained a new level of acceptance, if not respectability, with Donald J. Trump's unexpected election as President of the United States. The irony of the 2016 election is that, despite his "outsider" status, Trump's overwhelmingly White base stood to lose the most by voting for him. A vote for Trump also advanced rather than restrained racism.

Why is this? Why did votes for Trump represent a victory for White racism, and simultaneously, a defeat for White Americans? Certainly, not everyone who cast their ballot for him was a racist; to say that his opponent, Democrat Hillary Clinton, was an entirely polarizing candidate, dragging along more baggage than a freight train, does not overstate the case. Millions of people who cast their ballots for Trump were more interested in keeping Clinton out of the White House than they were in installing her opponent. However, it cannot be denied that part of Trump's appeal includes his racist, Islamophobic, xenophobic, and misogynist attitudes, which he took no pains to hide during the campaign.

Jeet Heer wrote in *The New Republic*: "Trump's essential appeal is based on racism. He launched his campaign talking about Mexican "rapists" and subsequently stirred up xenophobia against many other groups, especially Muslims. His racist pitch succeeded because the Republican Party is overwhelmingly White and has relied heavily on dog-whistle appeals to racism since the early 1960s. Trump appeals to the aggrieved privilege of well-to-do White Republicans who feel threatened by America's changing demographics and changes to the traditional racial hierarchy in the age of Obama (Heer 2016)." However, it was not simply the "well-to-do White Republicans" who felt threatened and voted for Trump. The not so well-off are even more enthusiastic in their support.

During the campaign, Trump not only made disparaging remarks about Mexicans; he criticized the judge presiding over a lawsuit filed by former students of his now-defunct Trump University. He said because the judge "happens to be, we believe, Mexican" (Gutman 2016), he was somehow biased against Trump because of his plans to build a wall on the Mexican border. The judge in question, Gonzalo P. Curiel, was born and raised in Indiana.

He also stereotyped African-Americans, saying, again during the campaign, that Black neighborhoods were "war zones," that African-Americans struggle to get by on food stamps and see nothing but failure all around them (Fausset et al. 2016). For the vast majority of African Americans, this is simply not true. By the end of the campaign, John McElwee wrote in *Dem*os, "It's increasingly difficult to deny the fact that racism played a key role in the 2016 presidential election. Donald Trump benefited from existing trends of polarization along racial attitudes and also propelled those trends forward" (Marcotte 2017).

Although Islamophobia is not, specifically, racism, it is included in this discussion because race theory has moved away from understandings of race, which were primarily concerned with biological explanations for differences between races. The concept of "racialization" allows scholars to examine the Muslim or Muslim American experience as racial. "We often hear that, because Muslims are not a race, people cannot be racist for attacking Muslims. This argument does not stack up. It is a simplistic way of thinking that overlooks the role that race plays in Islamophobic hate crimes" (Ruth 2018). According to Omi and Winant, "Racialization signifies the extension of racial meaning to a previously racially unclassified relationship, social practice or group" (Nye 2017). Furthermore, discrimination against Muslims by Trump's legions of fans is strong.

Furthermore, on December 5, 2016, Trump said this: "Donald J. Trump is calling for a total, and complete shutdown of Muslims entering the United States until our country's representatives can figure out what the hell is going on" (Lewis 2017). He did this to please his racist and xenophobic base, thoroughly strengthening cultural racism. Since becoming president, he has tried, with some success, not to "figure out what the hell is going on." Instead, he prevented Muslims from entering the United States. He supports the idea of a database to register all Muslims (Aziz 2017) and appointed retired General Michael Flynn as his first national security adviser. Although summarily dismissed after only three weeks on the job, Flynn drew scrutiny for anti-Muslim comments he has made over the years. In February, shortly after he left office, Flynn tweeted that "fear of Muslims is rational" ("Islamophobia Is Not a National Security Strategy" 2016).

Trump's racist attitudes and the policies that now accompany them were not mysterious; racism can be divided into three broad categories, cultural racism being one of them. We have seen how Trump fits within it; we will now look at the other two.

As a candidate, Trump could be described as what C.P Jones defines as an *Individual Racist.* Jones describes an *Individual Racist* as someone "... who believes that people of color are inferior to Whites because of physical traits, that these traits render their social, moral and intellectual capabilities inferior, and this is a legitimate reason for treating them as inferior" (Jones 1981, 27).

It is incredible to think that a racist candidate,—and based on Jones' definition, Trump certainly qualifies,—who made these statements and expressed these sentiments repeatedly was elected president of the United States. It is undoubtedly an indication of his primary opponent's weakness, the media's lack of attention to third-party candidates, and the Electoral College's oddity that Trump was elected.

Once elected and inaugurated, Trump became the embodiment of *Institutional Racism*, which Jones describes as "… established laws, customs, and practices which systematically reflect and produce racial inequality in American Society …" (Jones 1981, 28). His border wall, banning of travel from mostly Muslim countries, and praise for White supremacists all give credibility and some level of acceptability to these practices. This further encourages *Cultural Racism*, as described earlier. Many people with these attitudes voted for Trump. Doing so causes more damage to themselves and brought more challenges to them than any other election.

His presidential actions have only reinforced the view that he is racist. He initially appointed as Attorney General, former Senator Jeff Sessions (R-Ala.). In 1986, the Senate refused to confirm him as a federal judge due to comments he made suggesting he was racially prejudiced. This includes stating that he thought the Ku Klux Klan was "OK," until he learned that some of its members smoked marijuana (Sullum 2017). His nomination as attorney general was vehemently condemned by civil rights groups (Marans 2016) and praised by leading White nationalists, who are also pleased with Trump.

William Barr followed Sessions as Attorney General, who has harshly criticized the Black Lives Matter demonstrations that followed the killings of several unarmed and otherwise defenseless Black men (Norman 2020). Yet, he has been strangely silent about White terrorists, including those arrested for planning to kidnap Michigan Governor Gretchen Whitmer (Perez and Carrega 2020).

For Treasury Secretary, he appointed Steve Mnuchin, who has been accused of profiting from racial discrimination due to his racist practices as a hedge fund manager, and purchase of a troubled mortgage bank. The bank that Mnuchin ran "… has been hit with a new federal complaint of alleged redlining and other discrimination against minorities" ("Islamophobia Is Not a National Security Strategy" 2016). "Redlining" is described as "the unethical practice where financial institutions make it extremely difficult or impossible for residents of poor inner-city neighborhoods to borrow money. Also, they cannot gain approval for mortgages, take out insurance, or gain access to other financial services because of a history of high default rates in their communities. In this case, the rejection does not take the individual's qualifications and creditworthiness into account" (Kenton 2020).

In December of 2017, in the special election to fill the Senate seat vacated by Jeff Sessions, who became Attorney General, Judge Roy Moore was the Republican nominee. Trump endorsed him. CNN contributor Issac Bailey contends that they are kindred spirits because both support "… fewer rights for Americans who aren't White, male and straight, and more for those who cling to a perverted version of conservative Christianity…" (Bailey 2017). Moore was twice removed from the bench, first in 2003 for refusing to remove a statue of the Ten Commandments from the Alabama Judicial building, and then in 2016, after having been elected again, this time for ordering Alabama's probate judges to continue to deny marriage licenses to same-sex couples. A study by the Williams Institute of the UCLA School of Law indicates that 61% of the U.S. gay population is White (Institute 2020). While the number of gay men and women who vote Republican is probably minimal, many of their parents are Republican. This is one situation wherein Whites supporting a White racist are mainly penalizing other Whites.

Trump's many Executive Orders reflect racism. Order 13,768 is one example. This one, like many measures of immigration enforcement, was supposed to protect "public safety." "It begins with a strikingly broad statement that links immigration control to (undefined) ideals of national security and safety" (Kanstroom 2018). The words of the order reject "… any sort of discretionary, targeted, prioritized enforcement methods and aims…" (Kanstroom 2018). Anyone is targeted who is "illegal," regardless of how long they have lived in the country and if they have children born in the United States, or no matter how law-abiding they have been. This policy pleased Trump's base, who fear those who are not both White and Christian.

The White Power-Elite

To explore the impact of racism on Whites, understanding what is meant by the terms "elite" and "ruling class" is fundamental. To say that they are the "one percent" by income or wealth is too simplistic. The trajectory of income inequality is not a straight line with the rich always getting richer and the poor always getting poorer. There are times when, as President John F. Kennedy mentioned in 1963, "a rising tide lifts all boats" (Paik and Wiesner 2013). However, some of those are rowboats, and others are yachts, and distinguishing between the owners of the two is important.

According to Jill Lepore, "Income inequality is greater in the United States than in any other democracy in the developed world" (Lepore 2015). How has this happened, in a nation that prides itself on being the world's foremost democracy (exploring that myth is outside of the scope of this work)?

In C. Wright Mills' landmark book, *The Power Elite*, he said: "The power elite is composed of men (author's note: and women) whose positions enable them to transcend the ordinary environments of ordinary men and women; they are in positions to make decisions that have major consequences." Further: "… they are in command of the major hierarchies and organizations of modern society. They rule the big corporations. They run the machinery of the state and claim its prerogatives. They direct the military establishment. They occupy the strategic command posts of the social structure, in which are now centered the effective means of the power and the wealth and the celebrity which they enjoy" (Mills 1989).

Note that, according to Mills, the power elite is centered in three areas:

- Corporate—business and industry
- State—governance
- Military.

These three areas are increasingly combining forces. To again quote Mills: "…the economy is at once a permanent-war economy and a private-corporate economy. American capitalism is now in considerable part military capitalism, and the most important relation of the big corporation to the state rests on the coincidence of interests between military and corporate needs, as defined by warlords and the corporate rich" (Mills 1989, 275–76). Add to that Congressional rubber-stamping of nearly any military adventurism proposed. We can clearly see that the movers and shakers of U.S. policy are deeply entrenched in government, business, and the military. The lines between them continue to be blurred in the increasing size and power of what President Dwight D. Eisenhower referred to as the "Military-Industrial Complex."

Professor G. William Domhoff noted that "…policy-planning networks are primarily clusters of think tanks, philanthropic foundations and policy discussion groups that are supported by the upper class and corporate community – hereafter the ruling class - and that tend to generate consensus on broad policy issues that favor ruling-class interests" (2018).

Mills describes the "power elite," which is his term for those who pull strings. This is distinct from the "elite," the population that benefits from the directives of the "power elite," but is more or less content to sit back and let someone else do the work. He rejects the term ruling class because "Class is an economic term; 'rule' a political one. The phrase, 'ruling class,' thus contains the theory that an economic class rules politically" (Domhoff 2018, 277). Using "power elite" makes an important distinction.

Since the power elite members call the shots, they frame any and all policy discussions in ways that promote their interests and maintain their power The masses seem to have come to accept these self-benefitting proclamations as reality and do not question them. This is not always the case; witness the "Occupy Wall Street" movement, which, for a time, questioned the basic premise of the U.S.'s lords and masters.

It would be naïve to say that all members of the power elite think precisely the same. However, their goals are the same. They wish to organize the country and world in a way to maintain and increase their power and wealth. There are, of course, some differences of opinion about how best to achieve these unholy goals. However, "the power elite have a set of shared global values, such as the expansion of Western capitalism, that trump minor differences and serve to unite them" (Bonn 2010, 10).

What about Congress? What about that body of men and women who make the laws, often introducing legislation that has been written by powerful lobby groups, tools of the "power elite"? Alan Rappeport reported in the *New York Times* of January 12, 2015, that "The median net worth of a member of Congress was $1.03 million in 2013, compared with $56,355 for the average American household" (2015). The NYT report says that over half of the members of Congress are millionaires. New York Senator Charles Schumer (D) has said that members of Trump's cabinet have "more wealth than a third of the American people total – close to 100 million people" (Bay News 9 Staff 2017).

It cannot be denied that members of Congress legislate in their own best interests. For example, Trump's first Secretary of Health and Human Services was Tom Price. He is an orthopedic surgeon and former member of the House of Representatives from Georgia, whose net worth is estimated at $10 million. In 2016, while still a member of the House, Price bought stock shares in a medical device manufacturer called Zimmer Biomet. Within a week of this purchase, Congressman Price introduced legislation that directly benefited Zimmer (Raju 2017).

Other examples abound. "Rep. Mike Kelly, a Republican of Pennsylvania, was allowed to sponsor a natural gas bill the same month Exxon negotiated a multimillion-dollar deal with his wife. Across the aisle, former Rep. Dennis Cardozo, Democrat of California, helped pass a bill that involved tax breaks for racehorse owners and then bought seven such racehorses when the new tax breaks came into effect" (Abad-Santos 2013). Surprisingly, none of these individuals violated the letter of the law. "The practice is both legal and permitted under the ethics rules that Congress has written for itself, which allow lawmakers to take actions that benefit themselves or their families except when they are the lone beneficiaries" (Abad-Santos 2013). However, one wonders how important to Cardozo's constituency were tax breaks for people owning racehorses.

Many corporate bigwigs have been appointed to powerful positions by Trump. The following is just a small sample ("Trump's Corporate Cabinet" 2020):

- Interior Secretary David Bernhardt, who "has been focused on giving corporations greater access to natural lands and public resources throughout his professional career."
- Transportation Secretary Elaine Chao was secretary of Labor under George W. Bush. During that time, she "was deferential to businesses that mistreated their employees. This left workers vulnerable to wage-theft violations, as cases of employers illegally underpaying their workers continued to remain unchecked."
- Steve Mnuchin, Secretary of the Treasury. "At Treasury, delivering massive benefits to big corporations has been a top priority for Mnuchin."
- Environmental Protection Agency Administrator Andrew Wheeler is a former lobbyist for a coal producer.

In addition to appointing his corporate cronies to positions of power, Trump has surrounded himself with military men. On August 1, 2017, "…Attorney General Jeff Sessions named Gen. Mark S. Inch as the director of the Federal Bureau of Prisons (BOP), days after Gen. John Kelly moved from his position as secretary of Homeland Security to be White House Chief of Staff. Trump's Cabinet at one point contained three generals—Kelly along with Secretary of Defense, James Mattis, and Lieutenant Gen. Michael Flynn, who initially served as national security adviser" (Twitter 2017).

But these are political appointees; Congress members must stand for re-election, and should they have done something to displease their corporate or military masters, those in the power elite will simply send their sizable campaign contributions to a more malleable competitor. Congress remains interchangeable: the real "power elite" don't need to worry about their status or staying elected. They are completely secure in their positions.

That a few, mainly White, mainly male, people have an inordinate amount of control over all aspects of the lives of a significant number of the world's population can only be achieved with the help of a willing press. On June 5, 1991, the Bilderberg meeting in Sand, Germany, was held. Bilderberg is an elite conclave of banking, political, media, and industrial elites committed to world government. The subject of this particular address was media's role in promoting the power elite's objectives. David Rockefeller made this astonishing statement: "We are grateful to the *Washington Post*, *The New York Times*, *Time Magazine*, and other great publications whose directors have attended our meetings and respected their promises of discretion for almost forty years" (Grigg 2003). "'It would have been impossible for us to develop our plan for the world if we had been subject to the bright lights of publicity during these years,' continued Rockefeller. 'But the world is now more sophisticated and prepared to march towards a world government which will never again know war but only peace and prosperity for the whole of humanity'" (Grigg 2003). That this lofty prediction has not been fulfilled is obvious. But it is also obvious that trusting independent journalism to expose, let alone control, the power elite, is an exercise in futility.

It is clear to see that the wealthy are in charge of the government, with political appointees and the high number of millionaires elected to make laws, and that they make laws to their advantage. Their average net worth is over 18 times that of the average American. And what of race? The current Congress is 80% White, although racial minorities make up 38% of the nation's population (Bialik and Krogstad 2020). The result is that laws in the U.S. regulating domestic and foreign policy are still created mainly by wealthy Whites, overwhelming male (81%) (PolitiFact Staff 2017).

The Impact of Racism on Non-elite Whites

What does all this mean for non-elite Whites? How could discrimination against minorities not benefit them? How did cultural racism enable Trump and Congress to enact policies that benefitted them and their wealthy cronies, and few others?

As most Whites continue to support Trump and the White power elite, they pay the price. For example, in 2017, Vermont Sen. Bernie Sanders introduced an amendment that would urge the federal government to allow U.S. citizens to purchase pharmaceutical drugs from Canada. Such drugs are much cheaper there. A report from June 2019, about a "caravan" of people with diabetes going to Canada to get insulin, indicates that a vial of the life-saving substance in the United States costs $320.00; the same medication in Canada can be obtained for $30.00. A poll in 2015 indicated that 72% of U.S. citizens supported the importation of drugs from Canada. However, the measure was defeated. While some Democrats opposed the measure (notably

New Jersey Senator Cory Booker, who received more campaign contributions from the pharmaceutical industry over the previous six years than any other Democratic senator), the majority of Republicans voted against it (Jilani and Dayen 2017).

Republicans and conservatives praised the 2014 (pre-dating Trump) Supreme Court decision in Burwell v. Hobby Lobby Stores, Inc. In that case, the Court ruled that the government cannot require employers to provide insurance that includes birth control. The people who most need this service to be covered by insurance are those who cannot afford contraception by themselves. These are often Republicans who oppose government-sponsored or government-mandated health care, and Whites overwhelmingly identify as Republican compared to non-Whites.

Total student debt in the United States is estimated to be $1.45 trillion (StudentLoan Hero Staff 2020). However, while no demographics are available on the ethnicity of those owing the money, some information can be extrapolated from college graduates' demographics. "More than 37 percent of non-Hispanic White Americans have a college degree, while just 23 percent of African-Americans have reached the same level of formal education" (Wilson 2017). Since more Whites have college degrees than minority populations, it is safe to assume that more Whites are burdened by student debt.

The United States has an extensive public education program; most citizens send their children to public schools financed by local property taxes. According to Ranking America, the country ranks fourteenth in cognitive skills and educational attainment (RankingAmerica Staff 2020). With a White population of over 60%, it is mostly White children who are negatively impacted by cuts to education and Republican efforts to weaken teachers' unions.

In 2016, over 4 million White children lived in poverty in the United States (DataCenter Staff 2020). The same study indicates that about 3.5 million Black children were living in poverty at the same time in the country. These numbers are based on the U.S.'s official poverty line, which is far below what constitutes' actual poverty.

This plays out in other ways, too. Jeff Madrick states: "In 2010, Whites made up 64 percent of the population but received 69 percent of social benefits. Hispanics received only 12 percent of government benefits, though they made up 16 percent of the population. Included in this calculation are benefits from Social Security, Medicare, Medicaid, food stamps, TANF (Temporary Assistance for Needy Families), and housing assistance, and major tax credits such as the Earned Income Tax Credit and the Child Tax Credit."

"Conversely, when benefits are cut, they have a greater impact on White Americans than on those of color - a fact to which the electorate seems oblivious" (Madrick 2020, 103). Trump's budget proposals for 2021 included huge cuts to safety-net programs.

Combatting individual racism is not difficult; exposure to different peoples, cultures, and societies is often all it takes, but the individuals most needing that exposure are those most opposed to it. One example is instructive.

In October of 2015, an anti-Islam protest was scheduled in Ohio, near a mosque, to which only one person arrived. She spewed hatred, accusing Muslims of being terrorists, murderers, and Satanists. The counter-protesters tried to talk to her, to no avail, until one Muslim woman approached her and asked to hug her. The woman accepted first the hug and then an invitation to enter the mosque. When she left, she apologized and said, "I had no idea that people would be so nice to me, especially after me walking around with these signs" (Muslim Village Staff 2015).

Institutional racism was supposed to have ended with the Civil Rights measures signed into law by President Lyndon B. Johnson in 1964. These measures prohibited racial discrimination in public places; they integrated schools and outlawed employment discrimination based on race.

But no law can end cultural racism, although other laws can slowly bring about its end. For example, racial profiling is still performed in many places in the United States; countless young Black men have been killed for such crimes as walking while Black, riding in a car while Black, or shopping while Black. Their murderers, usually White police officers, are almost always exonerated. These shocking and blatant injustices must cease. The Black Lives Matter movement's resurgence following the shocking murder of George Floyd by a White police officer indicates a growing awareness among people of all color of institutional racism in the United States. However, as long as the power elite remain mostly White and the population blindly follows their directives, no substantive change can be expected.

Conclusion

It is not difficult to see how, once again, White Americans' tendency to elect Whites to the highest offices in the land does not bring them the benefits they crave. The election of Donald Trump brought that fact to the fore.

Many Whites see their way of life threatened by minorities, so they vote for like-thinking people, to their own detriment. The election to the presidency of Donald Trump was the crowning "achievement" of this twisted and unhealthy philosophy.

Following the 2012 re-election of President Barack Obama, FOX News anchor Bill O'Reilly said this: "Obama wins because it's not a traditional America anymore. The White establishment is the minority. People want things" (LoGiurato 2012).

Perhaps here we have the crux of the matter. O'Reilly, who can easily be seen to represent racist Whites, decries the fact that "the White establishment is the minority." That, for him, is something strange, alien, and frightening. It ends "traditional America." It causes people to "want things" from a government that is supposed to provide them with certain "things," including health care, quality basic education, affordable higher education, services for the poor, and decent infrastructure.

Based on the information included herein on poverty, student debt, and other topics, why should anyone but the White elite fear the end of "traditional America"? And with Trump's election, the racist, White-nationalist voting bloc has established itself as a force to be reckoned with, so any politicians seeking election will hesitate to criticize it too strongly. Regarding racism, it is business as usual in the United States.

Discussion Questions

Q1: Who are the White power elite? In what three areas is their power concentrated?

Q2: How much of Donald Trump's cabinet was filled by members of the White power elite?

Q3: What happened to consumer, environmental, educational, and financial regulations during the Trump presidency? How did this impact non-elite White Americans?

Q4: How have cuts to government spending on higher education impacted White Americans?

References

Abad-Santos, A. 2013. "Lucky for Congress, Blatant Conflict of Interest Is Still Perfectly Legal," October 30. https://www.theatlantic.com/politics/archive/2012/10/lucky-congress-blatant-conflict-interest-still-perfectly-legal/322790/.

Aziz, S. F. 2017. "A Muslim Registry: The Precursor to Internment?" *Brigham Young University Law Review* 4.

Bailey, I. 2017. "Why Trump and Moore Are Kindred Spirits," December 12. http://www.cnn.com/2017/12/12/opinions/roy-moore-donald-trump-amendments-states-rights-bailey-opinion/index.html.

Bay News 9 Staff. 2017. "PolitiFact: Trump Cabinet Wealthier Than a Third of America's Total Population?" *Bay News 9*, April 17. https://www.baynews9.com/fl/tampa/news/2017/4/18/politifact_trump_cab.

Bialik, K., and J. Krogstad. 2020. "115th Congress Sets New High for Racial, Ethnic Diversity," May 30. http://www.pewresearch.org/fact-tank/2017/01/24/115th-congress-sets-new-high-for-racial-ethnic-diversity/.

Bonn, Scott A. 2010. *Mass Deception: Moral Panic and the U.S. War on Iraq*. New Brunswick: Rutgers University Press.

DataCenter Staff. 2020. "Children in Poverty by Race and Ethnicity: KIDS COUNT Data Center." Accessed October 22. http://datacenter.kidscount.org/data/tables/44-children-in-poverty-by-race-and-ethnicity.

Domhoff, G. W. 2018. *Studying the Power Elite: Fifty Years of Who Rules America?* New York, NY: Routledge.

Fausset, R., Alan Blinder, and John Eligon. 2016. "Trump's Baffling View of Black America: Many Say Their Lives Bear Little Resemblance to His Portrayal in id for Votes." *The New York Times*, August 26. https://www.questia.com/read/1P2-39971403/trump-s-baffling-view-of-black-america-many-say.

Grigg, W. N. 2003. "Behind the Bias: Instead of Investigating and Exposing the Actions of the Power Elite, the Major Media Are Complicit in That Elite's Drive for Total Control." *The New American*, Februrary 10.

Gutman, David. 2016. "State GOP Silent as Trump Attacks Judges Ethnicity." *Charleston Gazette Mail*, June 4.

Heer, J. 2016. "Are Donald Trump's Supporters Idiots?" *New Republic*, May 11.

Institute, UCLA Williams. 2020. "LGBT Proportion of Population: United States." https://williamsinstitute.law.ucla.edu/visualization/lgbt-stats/?topic=LGBT.

"Islamophobia Is Not a National Security Strategy." 2016. *Tribune-Review/Pittsburgh Tribune-Review*, November 18.

Jilani, Z., and D. Dayen. 2017. "Cory Booker Joins Senate Republicans to Kill Measure to Import Cheaper Medicine From Canada," January 12. https://theintercept.com/2017/01/12/cory-booker-joins-senate-republicans-to-kill-measure-to-import-cheaper-medicine-from-canada/.

Jones, James M. 1981. "The Concept of Racism and Its Changing Reality." In *Impacts of Racism on White Americans*, edited by Benjamin P. Bowser and Raymond G. Hunt, 27–50. Beverly Hills, CA: Sage.
Kanstroom, D. 2018. "Expedited Removal and Due Process: 'A Testing Crucible of Basic Principle' in the Time of Trump." *Washington and Lee Law Review* 75 (3).
Kenton, W. 2020. "Redlining," October 20. http://www.investopedia.com/terms/r/redlining.asp.
Lepore, J. 2015. "Richer and Poorer." *The New Yorker*, March 16.
Lewis, H. 2017. "The Banality of Trumpism: Is the 'Muslim Ban' the Brainchild of a Chaotic Narcissist or a Budding Authoritarian?" *New Statesman*, February 2.
LoGiurato, B. 2012. "Bill O'Reilly Goes Off: 'The White Establishment Is the Minority'." *Business Insider*, November 6. http://www.businessinsider.com/bill-oreilly-election-obama-white-minority-establishment-romney-2012-11.
Madrick, J. G. 2020. *Invisible Americans: The Tragic Cost of Child Poverty*. New York: Knopf Doubleday.
Marans, D. 2016. "Civil Rights Groups Condemn Nomination of Jeff Sessions As Attorney General." *Huffington Post*, November 18. http://www.huffingtonpost.ca/entry/civil-rights-groups-condemn-jeff-sessions-appointment_us_582f12fde4b058ce7aaaa2b8.
Marcotte, A. 2017. "New Election Analysis: Yes, It Really Was Blatant Racism That Gave Us President Donald Trump." *Salon*, October 22. http://www.salon.com/2017/04/19/new-election-analysis-yes-it-really-was-blatant-racism-that-gave-us-president-donald-trump/.
Mills, C. W. 1989. *The Power Elite*. London: Oxford University Press.
Muslim Village Staff. 2015. "The Power of a Hug: Anti-Islam Protester Visits Mosque." *Muslim Village*, October 18. https://muslimvillage.com/2015/10/18/114088/power-hug-anti-islam-protester-visits-mosque/.
Norman, T. 2020. "The Dark Heart of Attorney General Bill Barr." *Pittsburgh Post-Gazette*, September 18.
Nye, M. 2017. "Race and the Study of Religion." https://medium.com/religion-bites/race-and-the-study-of-religion-53a43200b914.
Paik, P. Y., and M. E. Wiesner. 2013. *Debt: Ethics, the Environment, and the Economy*. Bloomington: Indiana University Press.
Perez, E., and C. Carrega. 2020. "Attorney General Barr Stays in Background After Michigan Arrests," October 10. https://www.cnn.com/2020/10/09/politics/barr-michigan/index.html.
PolitiFact Staff. 2017. "PolitiFact—Diversity in Congress: Democrats Have Women and Minorities, RepublicansHhave White Men Over 55?" April 28. http://www.politifact.com/wisconsin/statements/2017/apr/28/mark-pocan/congress-democrats-have-women-and-minorities-repub/.
Raju, M. 2017. "First on CNN: Trump's Cabinet Pick Invested in Company, Then Introduced a Bill to Help It." *CNN*, January 17. http://www.cnn.com/2017/01/16/politics/tom-price-bill-aiding-company/index.html.

RankingAmerica Staff. 2020. "Education." Accessed October 22. https://rankingamerica.wordpress.com/category/education/.

Rappeport, A. 2015. "Making It Rain: Members of Congress Are Mostly Millionaires." *The New York Times*, January 12. https://www.nytimes.com/politics/first-draft/2015/01/12/making-it-rain-members-of-congress-are-mostly-millionaires/?mcubz=0.

Ruth, D. 2018. "Indicting Islamophobia." *USA Today*, 39.

StudentLoan Hero Staff. 2020. "U.S. Student Loan Debt Statistics for 2020." Accessed October 20. https://studentloanhero.com/student-loan-debt-statistics/.

Sullum, Jacob. 2017. "Trump's Troubling Attorney General Pick." *Reason*, March, 2017. "Trump's Corporate Cabinet." 2020. Last Modified September 22, 2020. Accessed October 22. https://www.citizen.org/article/corporatecabinet/.

Twitter, E. 2017. "Military Figures Are Taking Over Trump's Administration." *Thinkprogess.org*, August 3. https://thinkprogress.org/trump-generals-policy-military-473c554cca72/.

Wilson, R. 2017. "Census: More Americans Have College Degrees Than Ever Before." *The Hill*. http://thehill.com/homenews/state-watch/326995-census-more-americans-have-college-degrees-than-ever-before.

6

Foreign Policy: A Double-Edged Sword—A History of Racism in U.S. Foreign Policy

Chris Danielson

Abstract U.S. domestic racism directly and profoundly influences the conduct of U.S. foreign affairs. Defending slavery and isolating foreign threats to it was the major priority of U.S. foreign affairs until the Civil War. Until the end of World War II, immigration policies played a major role in race relations and maintaining control over rapidly increasing industrial labor. The U.S.'s reluctant engagement in overseas colonialism in the last century was due to wanting to avoid making large numbers of non-White people citizens. Chris Danielson points out that a consistent issue in U.S. foreign policy is the desire of power elites in business, the military, and government to make their social class interests the same interests of the nation. One direct consequence is that non-elite Americans have fought and died in wars of questionable importance. Donald Trump's racism and nationalism were directly reflected in U.S. foreign policies toward all non-European nations. Politicians who follow Trump also view foreign affairs as a clash of civilizations—us versus them.

Keywords Foreign policy · Racism · Power elites · Trump · White Americans · Social class

C. Danielson (✉)
Montana Technological University, Butte, MT, USA
e-mail: CDanielson@mtech.edu

D. W. Austin and B. P. Bowser (eds.), *Impacts of Racism on White Americans In the Age of Trump*, https://doi.org/10.1007/978-3-030-75232-3_6

In the first and second editions of *Impacts*, Philip White and Gerald Horne reviewed race in foreign affairs. They pointed out that, as a British colony, the U.S. started as an outpost in British expansion of capitalism and colonization. From the colonies, the enslavement of Africans in the Americas and Caribbean financed European colonial expansion and the beginnings of the industrial revolution. Until the French lost the 1763 French and Indian War, Native Americans and enslaved Africans were potential allies against the British and slavery. Haiti's 1804 violent liberation from the French posed a direct threat to American slavery. In the War of 1812, some Native Americans allied again with the British, and enslaved Africans fled to the British. British abolition of slavery after 1834 threatened American slavery once again. From U.S. Independence until the 1860–1865 Civil War, U.S. foreign policy toward Great Britain, France, and Haiti was defined by the threats these countries posed to a small class of plantation slavers who dominated the U.S. government. Meanwhile, slavery impoverished most White Southerners. Few non-slave owning Whites had access to capital to purchase slaves and rise financially and socially in Southern society.

White and Horne's histories exposed the extent that U.S. foreign affairs defended plantation upper-class's economic interests until the Civil War. It took the Civil War to break their hold and end slavery at a tremendous cost to ordinary Whites. Slaveholders convinced poorer Whites to defend a way of life that only impoverished them. The War cost White Americans 580,000–710,000 dead on both sides. However, the long-lasting outcome of the War was the replacement of one elite, Southern planters, with another, Northern industrialists. Wealth would no longer be generated by slave labor but rather by wage labor. After the War, emancipated Blacks had the potential to dominate labor in critical industries and, of course, to make wage demands. Blacks knew how to use the vote to reduce economic exploitation after experiencing Black Reconstruction. So, European immigrants were used to offset this Black potential. After all, the Civil War only ended slavery, not racism. Industrialists learned to use Blacks as a reserve labor force to divide labor racially, producing conflict between Blacks and immigrants over jobs and housing. With Blacks as a reserve labor force, White labor could not demand higher wages, thereby improving owners' profits.

To say that race has played a role in American foreign policy would be an understatement. Yet, in U.S. foreign policy studies, the debate often traditionally focuses on realism versus idealism or the role of economic determinism. In recent years, scholars have given increased attention to how race and racism have affected foreign policy decision-making and the development of international capitalism, particularly the historical role of slavery in nineteenth-century economic development and the U.S.' industrialization. Race, along with gender, has become an increasingly important lens through which to analyze and interpret how White elites and policymakers shaped diplomacy and reacted to foreign events (Hunt 1987; Hoganson 2000; Scott 1986).[1] People of color and most White Americans have been exploited alike. This chapter's focus does not mean that race is the only factor, or even always the predominant factor in American diplomacy.

In many cases, race existed alongside other factors, such as geopolitics or economics. In some instances, it even worked on both sides, influencing an event's pros and cons. But race and who benefits from racial policies are still salient issues in American foreign policy today. From the earliest days of the republic, these policies worked to benefit White Americans' higher class at the expense of poor Whites and non-Whites alike. Such manipulation of interests continues even in the post-civil rights (but not post-racial) society of the early twenty-first century.

Earliest Policies

From its first days, the U.S. made racial considerations a major part of its foreign policy; slaveholders were its economic and political leaders. These leaders were primarily property owners from the White slaveholding class who sought to protect slavery as an economic institution that generated considerable wealth. This meant that Haiti, the Black republic produced by slave revolt, was isolated by the U.S. and would not receive U.S. diplomatic recognition until the Civil War (Dubois 2004, 225, 235, 303–4; Baptist 2014, 44–45; Forsdick and Hogsbjerg 2017, Chap. 6; James 1963). The U.S. attitude toward Haiti shows the complexity of racism in American foreign policymaking. Their objective was to prevent Haiti from spreading the "virus" of armed emancipation by enslaved Africans elsewhere in the hemisphere. However, the early policy did not wish to return Haiti to slavery since that would only strengthen its powerful French rival's economic and political position. These clashing interests—geopolitical concerns versus the

[1] For studies that emphasize economics as a decisionmaker in foreign policy, see William Appleman Williams, *The Tragedy of American Diplomacy* (New York: W.W.Norton & Co., 2009); *The Roots of the Modern American Empire* (New York: Random House, 1969), as well as Gabriel Kolko, *The Politics of War: The World and United States Foreign Policy, 1943–1945* (New York: Pantheon, 1990); Walter LaFeber, *The New Empire: An Interpretation of American Expansion 1860–1898* (Ithaca, NY: Cornell University Press, 1998). For realism and idealism, see Michael C. Williams, ed., *Realism Reconsidered: The Legacy of Hans J. Morgenthau in International Relations* (New York: Oxford University Press, 2008); Matthew C. Price, *The Wilsonian Persuasion in American Foreign Policy* (Amherst, NY: Cambria Press, 2007).

maintenance of White supremacy—would also carry over into other foreign policy matters in the nineteenth century. Racial concerns would also play a growing role in diplomacy as slavery expanded and became the leading issue in the republic's domestic politics leading up to the Civil War (Thomas 1997, 551–52, 614–16; Deyle 2006).

The 1840s and 1850s for the U.S. was the era of the "foreign policy of slavery," which included acquiring new lands for slavery, mainly from Indians and "mongrel" Mexicans (Karp 2016, 5–7).[2] When the Anglo-Saxon Texas settlers established an independent republic in 1836, expansionists sought to add the region as a slave state. However, the slaveholding class's desire to spread slavery clashed with the growing abolitionist movement in the U.S.. Thus, White slaveholding lawmakers decided Texas was not worth the risk of upsetting the Missouri Compromise's balance of slave and free states (Baptist 2014, 266–68, 297).[3] Yet geopolitical concerns soon countered this sentiment, as fears of British influence in Texas worked to push the Tyler administration to embrace annexation as an check on British expansion (Haynes and Morris 1997).

The desire for more land for slavery caused the Mexican War in 1846. Economic gain and racial exploitation motivated the War (Greenburg 2012, 57, 95–96, 102–4, 120, 131–35). The conquest of Mexico opened a debate over how much of Mexico to annex to the U.S. Some Democrats in Congress pushed for the entire country's annexation to the U.S., yet their racism worked both ways. Opponents of the "All Mexico" policy of the expansionists used Mexico's racial diversity to resist absorbing the entire country out of fears of making U.S. citizens of thousands of non-White Mexicans. Ultimately, concern over granting U.S. citizenship to Mexicans acted as a check on the U.S.'s appetite for the entire country. Racism motivated anti-imperialism *and* imperialism (Greenburg 2012, 212–13, 215–17).

The White supremacy of U.S. political and economic leaders had a global reach by the end of the nineteenth century. All the major European powers and the U.S. established dominions worldwide over people of color by force. After its defeat of Spain in 1898, the U.S. added the Spanish Caribbean and Pacific colonies to its empire and effectively turned the biggest prize of all, Cuba, into a vassal state. Yet as with Mexico, racism acted as a check on expansionism. Some leaders worried that the U.S. should not annex any territory in "the tropics," which again meant absorbing more dark-skinned

[2] Baptist, 154–58; Horsman, 208–212, 219–20; Matthew Karp, *This Vast Southern Empire: Slaveholders at the Helm of American Foreign Policy* (Cambridge: Harvard University Press, 2016), 5–7.

[3] Baptist, 266–68, 297.

people into the nation from climates believed unsuited for White people. These sentiments led the Senate, swayed by these racialized views of climate and its own racism, to deny the Grant administration's request to annex the Dominican Republic (Love 2004, 28, 33, 56, 59).

What effect did these early policies have on the vast majority of White Americans who neither owned slaves nor had a say in defining U.S. foreign policy? The following are a few historical effects:

1. In defending slavery, these foreign policies helped maintain the economic dependence of most White Americans on slavery in two ways. First, slavery promoted the Industrial Revolution by sending cotton to northern factories and creating new industrial jobs. Cotton sales created wage dependency for northern White workers. Second, Northern capital flowed south and promoted the expansion of slaveholding for White elites, but not for poorer Whites. Progressively fewer non-slave owning Whites could access the capital to purchase slaves and rise financially and socially in southern society.
2. The use of racial supremacy in U.S. foreign policy helped to obscure conflicting class interests among Whites and maintain unchallenged elite dominance of government and the national economy. White slaveholders played to poorer Whites' racial and economic fears, especially after the adoption of universal White male suffrage. White slaveholders maintained power by convincing poorer Whites that emancipation of slaves would mean competition with Black labor. There was also the threat to White supremacy of potential sexual access of Black men to White women.
3. Early foreign policies helped create a national identity based upon race and White racial superiority. Only Whites could be American and being White was superior to all others no matter how poor one was.
4. Early U.S. foreign policy helped to make poorer Whites willing foot soldiers of elite White economic interests. They put down rebellions by slaves and poor Whites, exterminated Native Americans for their lands, and sent poor Whites to fight in U.S. overseas adventures.
5. As the numbers of European immigrants swelled the ranks of landless Whites, so did the threat of violence against White elites. It was necessary to continually add new lands for speculation and settlement to maintain the White elites' social order. Continuing elite control of the government and the economy required an expansionist foreign policy.
6. What the poorer White majority gained from a racist social identity was the illusion that they were racially superior to enslaved Africans, Native

peoples, European immigrants, Haitians, Mexicans, Cubans, or any other foreigners their "masters" designated as inferior.

Asia and the Pacific

Racism as a factor in U.S. foreign policy was not just a matter of Black and White, especially after slavery. The growth of international trade and capitalism in the nineteenth century meant that the U.S. government promoted an economic system of trade geared toward enriching White capitalists at home and protecting White labor from non-White competition. U.S. economic expansion into the Pacific brought the question of Asia and Asians increasingly into the country's decision-making about immigration. The Naturalization Act of 1790 limited citizenship of new immigrants to Whites. In 1882, Congress passed the Chinese Exclusion Act, the first time the U.S. government banned a specific racial group from immigrating to the U.S. (Takakai 1998, 110–12, 114; U.S. Government 1882).

While China was a natural market for American goods, Japan drew far less interest. Nevertheless, Commodore Matthew Perry led a naval expedition to Japan in 1853 and intimidated the Japanese to sign the Treaty of Kanagawa, establishing diplomatic relations between the two countries. The opening of Japan had a profound effect on the destinies of both countries. Japanese officials consciously pursued a policy to industrialize their country by learning from the West, so they would not have to surrender to it (LaFeber 1998, 8, 10–17, 25–26, 29–30, 34–36).

Of far more strategic and economic interest to the U.S. was the Kingdom of Hawai'i, which was important as a naval base. By the 1880s, American sugar planters dominated the economy and reduced the Hawaiian monarch to a puppet government. In 1891, White planters, with U.S. help, overthrew Queen Lili'uokalani and stripped non-Whites of their political rights. The seizure of power was cruel enough for President Grover Cleveland to withdraw a treaty of annexation for Hawai'i, but geopolitical desires won out, and the islands were annexed in 1898. Indeed, Whites in Hawai'i used racism to counter the fears of other Whites, who were against adding non-Whites to the country as citizens, argued instead that annexation would protect American settlers from non-Whites (Okihiro 2015).

The U.S. did not engage in a large-scale war with Asians until the Spanish-American War. The U.S. occupation of the Spanish Philippines prompted a guerilla war by the Filipinos. Geopolitical and economic factors were part of the desire to keep the Philippines because of their proximity to China.

To overcome fears among anti-imperialists of the Filipinos being given U.S. citizenship, the Senate approved annexation with amendments denying citizenship to Filipinos; the treaty passed the Senate by only two votes (Kramer 2006, 1–2, 91).

Japan and World War II

Yet as the U.S.'s economic and geopolitical interest grew in the twentieth century, the U.S. found itself not just competing with White European nations but now a powerful non-White one as well. Having a U.S. overseas empire meant collision with the rising power of Japan. The Japanese desire for economic self-sufficiency threatened the economic privileges of White industrialists seeking Asian markets while White workers at home faced competition from Japanese immigrants. Westerners grudgingly acknowledged Japan's emergence, where it was held up as "civilized Asia." Racial dimensions increasingly shaped U.S.-Japanese relations, as Japanese immigration to the U.S. led to the rising anti-Japanese sentiment among Whites. The "Gentlemen's Agreement" of 1907 dealt with this temporarily, but each country had conflicting economic interests (LaFeber 1998, 88–90). Japan's desire for a sphere of influence opposed the U.S.'s desire to maintain an East Asia open to free trade. The U.S. and Britain also rejected Japan's proposed racial equality clause among nations at Versailles in 1919 (LaFeber 1998, 121–24).

Although both the U.S. and Japan agreed to naval disarmament in the Five-Power-Treaty in 1922, racial issues continued to poison U.S.-Japanese relations. In 1924, the Reed-Johnson Act banned Japanese immigration. This law's discriminatory effect undermined peaceful diplomacy and encouraged Japanese militant ultranationalists who saw the U.S. as a threat to Japan. Thus, U.S. racism played a part in the coming clash of the two empires in World War II (LaFeber 1998, 137–41, 144–46).

Perhaps, this worsening relationship could have been avoided. Supporting the Racial Equality Clause at Versailles could have headed off the immigration ban and continued diplomatic endeavors. Of course, the Great Depression may still have prompted the Japanese military's actions in China. More fundamentally, the U.S.'s acceptance of this clause would have profound implications for its domestic race relations. It would have emboldened civil rights activists at home to use the clause to agitate against White supremacy. These civil rights reformers could have gained international support for their actions. A similar challenge to British racial superiority would have occurred in British colonies like India, which was already experiencing independence agitation. For those reasons alone, U.S. and British policymakers never accepted the Racial Equality Clause at Versailles.

The Great Depression led to Japan's invasion of China, threatening the Open Door. Pearl Harbor's attack resulted from these two competing visions, one of an enclosed self-sufficient Japanese empire versus an Asia open to U.S. trade. It was also a war between a White and a non-White power, with all the racial baggage included (LaFeber 1998, Chap. 6).

U.S. racial prejudice against the Japanese during World War II was substantial and triggered a racist backlash not replicated in the War against Germany and Italy. President Roosevelt's Executive Order 9066, which interned without trial over 100,000 Japanese and Japanese Americans, was a decision not extended to German Americans or Italian Americans. Anti-Japanese racism also played a role in swaying President Harry S. Truman to use the atomic bombs, but racism was secondary to U.S. desires to intimidate the Soviets and shape the postwar world toward U.S. interests (Takaki 1995; Dower 1986; Daniels 1993; Office of the President 1942).

The racial change created by World War II had to do with two factors. The first was Japan's early successes over White imperialists, which shattered the aura of superiority Whites enjoyed over non-White peoples. The second was the Holocaust, which showed the full extent of state enforcement of racial inferiority and underminded offical racism. Japan's war successes forced American policymakers to support an antiracist movement in international relations, and gave rise to postwar antcolonial movements (Takakai 1998, 376–78; Jansen and Osterhammel 2017, 68–69, 82).

The Cold War and Third World Independence

The post-World War II period marked a major break from earlier eras. Race became even more salient as large numbers of non-White nations entered the world arena due to colonial independence movements. This heightened White anxieties, as now non-White nations would increasingly have to be dealt with as independent actors, not White-controlled dependent powers. Decolonization created dozens of new nations freed from White paternalistic political control. Also, the movement for African-American civil rights effectively discredited racism as a diplomatic action of nation-states. Previously, all the major European and North American nation-states were overtly racist, so whether one promoted a White supremacist agenda in foreign policy did not matter. Now, it did, and the moral credibility of the U.S. was on the line.

Racial considerations did not vanish, however. Now, a nation like the U.S. had to appear nonracist since public opinion mattered more in international relations than it had before. The Cold War rivalry with the Soviet Union meant that U.S. racial discrimination was thrust onto a global stage. Every act of racial violence harmed U.S. diplomats' efforts to make friends in the Third World (Dudziak 2000).

Again, racial considerations ran up against geopolitical and strategic concerns, which created complex dilemmas. Decolonization immediately after the War became enmeshed in Cold War security considerations for Europe. Here the relative importance of an ally mattered. Washington could pressure a small ally such as the Netherlands to cease fighting in the Dutch East Indies, which led to Indonesia's Republic. However, bigger, strategic allies like France—critical as a bulwark against Soviet influence—had more clout. Despite his misgivings, Truman acquiesced to Charles de Gaulle's desire to reclaim French Indochina, which led to the eventual U.S. underwriting of the French War. Thus, Washington's reluctance to commit itself fully to decolonization contributed to its involvement in the Vietnam War (Herring 1996, 10–11, 14–17, 21–26; Brands 1993, 32–34).[4]

In Africa, the U.S. took little interest in decolonization, but racial discrimination against African diplomats in Washington, D.C. led them to regard the U.S. as a hardship post. This gave President Kennedy a further incentive to push civil rights legislation, to improve relations with Africa (Nwaubani 2001, xv–xvi, 117–18, 160–61). The U.S. would now have to combat racism actively to help its relations abroad, but Black-versus-White violence in Africa created contradictions. Under Eisenhower and Kennedy, the U.S. supported White colonists in the Belgian Congo and Angola. The U.S. desired critical minerals in the former and in the latter access to strategic airbases in the Portuguese Azores (Borstelmann 2001, 128–29, 148–50). These events showed that the U.S. undermined its rhetoric regarding equality, especially with its support for Third World dictators. In non-White Third World countries, democracy was fine if pro-capitalist and pro-American regimes emerged. However, that was not often the case. A large country like India could pursue a nonaligned policy and resist Washington's wrath, but smaller countries did not enjoy such immunity. One only needs to look at Iran and Guatemala, two

[4] Jansen and Osterhammel, 83–84, 86. For differing perspectives on the Cold War and its effects on the early civil rights movement, see Robert Korstad and Nelson Lichtenstein, "Opportunities Found and Lost: Labor, Radicals, and the Early Civil Rights Movement," *Journal of American History*, Vol. 75, No. 3 (Dec. 1988), 786–811; Manfred Berg, "Black Civil Rights and Liberal Anticommunism: The NAACP in the Early Cold War," *Journal of American History*, Vol. 94, No. 1 (Jan. 2007), 75–96. For Indonesian independence, see Robert J. McMahon, *Colonialism and the Cold War: The United States and Indonesia Independence, 1945–49* (Ithaca, NY: Cornell University Press, 1981).

countries that in the early 1950s saw their elected governments overthrown in favor of U.S.-backed dictators who protected U.S. corporate property (Brands 1993, 57–58, 131–33).

The gap between American rhetoric for freedom and its actual foreign policy was greatest in its dealings with South Africa. Its strategic importance, particularly its uranium, offset Truman's unease with South Africa's 1948 adoption of apartheid. As Thomas Borstelmann put it, "[i]n swallowing apartheid, the Truman administration helped lay the groundwork for decades of American policy toward the Third World and also for the anti-American sentiment so evident in many of those regions today" (Borstelmann 1993, 106–7, 193, 203). The 1965 declaration of independence by Prime Minister Ian Smith of Rhodesia created a White supremacist bastion in Africa that, along with South Africa, drew support from White supremacist politicians like Senator James Eastland of Mississippi. The difference in how the U.S. treated White supremacist regimes versus non-White communist regimes was exemplified in 1966 when Secretary of State Dean Rusk said that "the thought of using U.S. troops [in Rhodesia] is appalling." Yet, at the same time, thousands were being used in Vietnam to kill Asians.

Rhodesia, pressured in part by President Jimmy Carter's economic sanctions, held democratic elections that established Zimbabwe's Black-ruled country in 1980. This showed how the civil rights movement had forced U.S. policymakers to change. But Carter's successor Ronald Reagan pursued a policy of constructive engagement with South Africa that did little to erode apartheid. Then, Congress passed economic sanctions against South Africa over Reagan's veto in 1986, and free elections finally came to South Africa in 1994 (Borstelmann 1993, 126–28, 154–55, 195–202; Asch 2008, 262–66).[5]

By the end of the Cold War, domestic racism had more direct effects on U.S. foreign policies affecting White Americans and people of color alike. Regardless of race, poor Americans paid the price for continued White supremacy and exploitation of people of color overseas. A few of those costs were:

1. An underlying cause of World War I and II was competition among the U.S. and European colonial nations for overseas colonies in Africa, the Middle East, Asia, and the Caribbean. The White poor and working-class

[5] See also Nancy Mitchell, *Jimmy Carter in Africa: Race and the Cold War* (Washington, DC: Woodrow Wilson Press, 2016), 7–8, 11–13, 15, 673–74; J.E. Davies, *Constructive Engagement?: Chester Crocker & American Policy in South Africa, Namibia & Angola 1981–8* (Athens, Ohio: Ohio University Press, 2007), 209–12. For more on Cuba's foreign policy in Africa, see Piero Gleijeses, *Visions of Freedom: Havana, Washington, Pretoria and the Struggle for Southern Africa* (Chapel Hill: University of North Carolina Press, 2013).

paid the heaviest price in deaths and casualties in both wars. Racist attitudes and U.S. foreign affairs toward Japan contributed to the War in the Pacific and additional deaths and casualties.

2. For the first half of the twentieth century, American labor's belief in White supremacy meant many unions excluded people of color. By refusing to organize Blacks and Latinos into the labor movement, labor unions unwittingly helped entire industries shift well-paid union jobs to non-union labor in the U.S. South and Latin America. This strategic error is a central factor in the current decline and waning influence of the U.S. labor movement.
3. Southern legislators responsible for racist U.S. foreign policies are also proponents of domestic policies that maintain poverty in the U.S. and retard modern social welfare in the U.S. To this day, they have consistently opposed improvements in healthcare, education, wages, environmental quality, workers' and voting rights for fear that African Americans might benefit. Their efforts to keep Blacks poor and subordinate have harmed far more Whites.

End of the Cold War and the Neocolonial Period

The collapse of the Soviet Union in 1991 ended the bipolar system of competition for influence in the Third World. Many in the West celebrated the end of Soviet communism and the expectation that free-market capitalism and liberal democracy would become the norm. However, in fighting the Cold War, non-Whites, as victims or proxies, suffered the greatest from the two superpowers' struggle. In southern and northeastern Africa, in Korea and Vietnam, and South Asia and Central America, eventually, millions died in civil wars internationalized by the global conflict (LaFeber 1993; Woodroofe 2013; Bass 2013; Fukuyama 1992).

Perhaps this all could have been avoided, with the U.S. respecting the self-determination of Third World nations. But to allow other countries to manage their affairs would have required U.S. policymakers to deviate from their ideological lens laid in the early Cold War and developed in the Jim Crow South. That blinkered vision inhibited any direct U.S.-Chinese relations for more than twenty years and was only changed due to China's strategic and geopolitical importance. An earlier, closer relationship with Beijing may have averted the Korean and Vietnam Wars (Kahn 1975; Joiner 2009; Service 1971).

By the 1990s, racism became reflected in more subtle ways. Having "won" the Cold War, the U.S. continued to exert its economic influence over poor, non-White countries through globalization. American consumers reaped the benefits, purchasing products produced by cheap foreign labor. Relatively inexpensive consumer goods helped mask the effects of the decline of U.S. industrial jobs due to globalization, which hurt the U.S. White middle and working classes. The U.S.'s support for Israel's occupation of Palestine has opened itself to charges that is supporting a new form of apartheid. Then, the September 11 attacks justified the military invasion of Iraq and Afghanistan. While ostensibly against Islamic terrorists, these conflicts have a racial effect as they reinforced racist attitudes against Arabs and Middle Easterners. The perils of climate change even have a racial aspect. Residents of wealthier, White countries, fearing a decline in their living standard, oppose making concessions to non-White, developing countries (Grandin 2007; Sides and Gross 2013; Takei 2012).

Trump

As the twenty-first century opened, it was unfashionable to display racism overtly. Donald Trump's election has changed all that, as he harnessed White anxieties about declining economic status in an era of globalization and job outsourcing. The depth of White working- and middle-class anxiety and alienation has surprised even domestic and international elites and is a rebellion against their control of government and economy. White supremacy is enjoying a resurgence, thanks to Trump's open prejudice against Mexicans and Muslims during the campaign. A resurgence in domestic racism from his political base is also why he refused to condemn neo-Nazis at a rally in Charlottesville, Virginia. Trump's refusal led to a rebuke from the United Nations Committee on the Elimination of Racial Discrimination (Lopez 2014; Chan and Cumming-Bruce 2007; Saletan 2017).

In foreign policy, the first months of Trump's presidency suggest he echoed Samuel Huntington's "Clash of Civilizations" thesis. After the Cold War, Huntington predicted that future conflicts would be culture-driven, superseding old nation-state conflicts. Reflecting the racial attitudes mentioned earlier on climate change, Trump withdrew the U.S. from the Paris Climate Accord in June 2017. However, his real embrace of Huntington was on display during his Warsaw speech. He said that "[t]he fundamental question of our time is whether the West has the will to survive," painting a threat of Muslim terrorism and immigration to Europe and the U.S. and echoing the themes of cultural war. Trump's rhetoric of the West in peril, even as it faces an enemy far less destructive than Nazi Germany, stands in sharp contrast to the two visits to Poland made by George W. Bush. Bush never referred to "the West" and instead discussed themes of universal values and world cooperation rather than cultural identity. Bush referenced democracy thirteen times, while Trump mentioned it only once (Huntington 1993; Beinart 2017). In Trump's rhetoric, the U.S. is under siege from racial and religious "Others."

Yet the largest and most important "Other" is China. With his hostility to free trade pacts, President Trump has effectively ceded global economic leadership to China, as well as strategic leadership. China has stepped into both U.S.-North Korean tensions and the Middle East peace process, forged trade links in Africa and Southeast Asia (including arms sales), and claims sovereignty over the South China Sea China with little U.S. opposition (Rapoza 2017; Martina 2017; Hayton 2017).[6] As China fills the void, a new "Yellow Peril" seems to be emerging as U.S. interests retreat before China's growing influence. Trump regularly scapegoated China, most recently calling COVID-19 a "Chinese" virus (Leigh 2020).

Trump's expressions of racism are strongest when it comes to immigration policy and dealings with non-White countries. He consistently raised during and after the 2016 election fears of an "invasion" of migrants from Central America and Mexico across the southern border. First, he issued executive orders banning immigrants from several predominately Muslim countries, which eventually expanded to include countries with Muslim minorities. He also ordered the construction of a border wall on the U.S.-Mexico border to stop migrants from entering the U.S., with plans to have much of the wall in place by Election Day 2020. Related to the wall was the separation of families detained at the border. Taking minor children away from their parents was so controversial that the public backlash led to the policy suspension in the summer of 2018. Much of this latter policy was due to advisors like Stephen Miller, a senior policy advisor with White nationalist viewpoints. In January 2018, during a meeting on immigration policy, Trump also referred

---

[6] See as well "China Competes with US for Weapons Sales to Thailand," *Bangkok Post*, 3 December 2019, https://www.bangkokpost.com/thailand/general/1807664/china-competes-with-us-for-weapons-sales-to-thailand.

Michael Martina, *Reuters*, 11 January 2015, "China Will Not Take Path of 'Western Colonists' in Africa: Foreign Minister," http://www.reuters.com/article/us-china-africa-idUSKBN0KL09720150112; Yun Sun, "Africa in China's Foreign Policy," John L. Thornton China Center and Africa Growth Initiative, Brookings Institution, April 2014, 2. For more on China and its investment in Africa, see Denis M. Tull, "China's Engagement with Africa: Scope, Significance and Consequences," in Christopher J. Lee, ed., *Making A World After Empire: The Bandung Moment and its Political Afterlives* (Athens, OH: Ohio University Press, 2010), 289–314.

to Haiti, El Salvador, and all of Africa as a collection of "shithole countries," an insult that drew international condemnation. The racial link was clear, as he also expressed a desire for "more people from Norway," one of the world's wealthiest and Whitest countries (Fritzw 2019; Kight and Kranz 2020).[7]

His constant raising of these issues, either through Twitter or at rallies, is intended to extend and maintain his reactionary White voting bloc. This bloc is particularly critical since it was key to his 2016 victory, as he won with a combination of predominately White, low-populated rural states, as well as disaffected White voters in key Democratic states. Despite losing the popular vote by almost three million votes, these voters allowed him to win in the Electoral College, the largest margin of losing the popular vote in history for a successful presidential candidate. Simply put, Trump used wedge issues to rally his base of many White Americans fueled by racial anxieties, many of whom are downwardly mobile. The effectiveness of these appeals has to do with the rise of White identity politics, which Trump harnessed in 2016. The growing interdependence of the U.S. in a globalized world has played a major role in shaping White anxieties about declining status nationally and internationally. A consequence of these trends and Trump policies is that White Americans, who have the most to lose from his policies, have lent support to Trump because they believe that his policies will preserve their living standards (McCormack 2016; Bump 2016; Jardin 2019).

Trump's racism and that of his White political base have had an immediate negative effect on both U.S. domestic and foreign policies. Some of those effects are:

1. Because of their xenophobia and racism, Trump and his base are hostile to globalism, which has become the basis of U.S. economic growth and continuity of the American middle class increasingly. Anti-global stances

[7] In addition, see Jessica Kutz, "Triumph and Tragedy: Trump's Border Wall Expands," *High Country News*, 20 April 2020, https://www.hcn.org/issues/52.5/south-borderlands-triumph-and-tragedy-trumps-border-wall-expands; Roger Sollenberger, "Miller's Hardline Refugee Policies Revived: 'Only a matter of time before the situation blows up'," 17 April 2020, https://www.salon.com/2020/04/17/millers-hardline-refugee-policies-revived-only-a-matter-of-time-before-the-situation-blows-up/; Michael Edison Hayden, "Stephen Miller's Affinity for White Nationalism Revealed in Leaked Emails," 12 November 2019, https://www.splcenter.org/hatewatch/2019/11/12/stephen-millers-affinity-white-nationalism-revealed-leaked-emails; "'Irresponsible Reprehensible, and Racist': Here's How the World Reacted to Trump's 'Shithole' Comments," *Business Insider*, 14 January 2018, https://www.businessinsider.com/reactions-around-the-world-to-trump-shithole-comments-2018-1; Terje Solsvik and Camilla Knudsen, "'Thanks, but No Thanks'—Norwegians Reject Trump's Offer," https://www.reuters.com/article/us-usa-trump-immigration-norway/thanks-but-no-thanks-norwegians-reject-trumps-immigration-offer-idUSKBN1F11QK. For studies of Trump's immigration policies, see Julie Davis Hirschfield and Michael D. Shear, *Border Wars: Inside Trump's Assault on Immigration* (New York: Simon & Schuster, 2019); Shoba Sivaprasas Wadhia, *Banned: Immigration Enforcement in the Age of Trump* (New York: New York University Press, 2019).

impoverish American farmers, close American overseas markets, and joint-business ventures to the detriment of all Americans.

2. Under Trump, the wealthiest and top one percent of White Americans continued to accumulate larger shares of the national wealth while the working- and middle-class shares declined. Sizeable portions of the White working and middle class support the very policies that impoverish them out of fear of White racial loss.
3. Again, White Southern anti-big government beliefs exist historically to oppose equal rights and economic opportunities for African Americans. This belief has been used under Trump to immobilize the Federal government, eliminate civil rights and social welfare policies that would reduce economic inequality and, in fact, benefit the majority of White Americans.
4. Most world governments represent people of color, who are 84% of the world's population. For Trump and the new Republican base, domestic racism carried over into foreign affairs. Allies and international rivals alike have been racially and culturally insulted and demeaned. The consequence is a questioning of the stability and longevity of U.S. global leadership, influence, and alliances.

Conclusion

From the nation's beginnings, racism has consistently played a role, sometimes overt and sometimes subtle, in U.S. domestic and foreign policymaking history. Racism has variously co-existed, competed with, or been subordinated to geopolitical or other factors. But does that have to be the case? Can Americans accept a foreign policy based on something other than neocolonialism and military interventions in non-White countries? It seems unlikely, as Trump quickly backed off his nonintervention stance by continuing the War in Afghanistan and Syria and covert operations in Niger. It would also require Americans to accept the right of developing countries to enforce their worker protections, which would raise the prices of cheap imports into the U.S. One only needs to look at the opposition of many middle-class Americans to an increased minimum wage in the U.S. as evidence of the deep-seated class hostility toward the working poor, especially workers of color. Perhaps it will take another global crisis on par with the Great Depression, such as the ongoing COVID-19 crisis, to reshape many White Americans' attitudes toward a more egalitarian and economically fair global society. Or perhaps nothing ever will.

Discussion Questions

Q1: In what ways does U.S. domestic racism influence the conduct of U.S. foreign affairs?

Q2: Prior to the U.S. Civil War, how did slavery of Africans and their descendants contribute to the poverty of most Whites in the Southern states?

Q3: After the U.S. Civil War, how did industrial elites use racism against free Blacks and immigrants to drive down wages for working-class Whites?

Q4: How did Donald Trump's opposition to globalism impoverish a generation of American farmers?

References

Asch, Christopher Myers. 2008. *The Senator and the Sharecropper: The Freedom Struggles of James O. Eastland and Fannie Lou Hamer*. New York: The New Press.

Baptist, Edward E. 2014. *The Half Has Never Been Told: Slavery and the Making of American Capitalism*. New York: Basic Books.

Bass, Gary J. 2013. *The Blood Telegram: Nixon, Kissinger, and a Forgotten Genocide*. New York: Alfred A. Knopf.

Beinart, Peter. 2017. "The Racial and Religious Paranoia of Trump's Warsaw Speech." *The Atlantic*. Accessed August 29. https://www.theatlantic.com/international/archive/2017/07/trump-speech-poland/532866/.

Borstelmann, Thomas. 1993. *Apartheid's Reluctant Uncle: The United States and Southern Africa in the Early Cold War*. New York: Oxford University Press.

———. 2001. *The Cold War and the Color Line: American Race Relations in the Global Arena*. Cambridge, MA: Harvard University Press.

Brands, H.W. 1993. *The Devil We Knew: Americans and the Cold War*. New York: Oxford University Press.

Bump, Philip. 2016. "Donald Trump Will Be President Thanks to 80,000 People in Three States." *Washington Post*, December 1. https://www.washingtonpost.com/news/the-fix/wp/2016/12/01/donald-trump-will-be-president-thanks-to-80000-people-in-three-states/.

Chan, Sewall, and Nick Cumming-Bruce. 2007. "U.N. Panel Condemns Trump's Response to Charlottesville Violence." *New York Times*, August 23. https://www.nytimes.com/2017/08/23/world/un-trump-racism-charlottesville.html?mcubz=3.

Daniels, Roger. 1993. *Prisoners Without Trial: Japanese Americans in World War II*. New York: Hill and Wang.

Deyle, Steven. 2006. *Carry Me Back: The Domestic Slave Trade in American Life*. New York: Oxford University Press.

Dower, John W. 1986. *War Without Mercy: Race and Power in the Pacific War*. New York: Pantheon Books.

Dubois, Laurent. 2004. *Avengers of the New World: The Story of the Haitian Revolution*. Cambridge: Harvard University Press.

Dudziak, Mary L. 2000. *Cold War Civil Rights: Race and the Image of American Democracy*. Princeton: Princeton University Press.

Forsdick, Charles, and Christian Hogsbjerg. 2017. *Toussaint L'Ouverture: A Black Jacobin in the Age of Revolutions*. London: Pluto Press.

Fritzw, John. 2019. "Trump Used Words Like 'Invasion' and 'Killer' to Discuss Immigrants at Rallies 500 Times." *USA TODAY Analysis*, August 8. https://www.usatoday.com/story/news/politics/elections/2019/08/08/trump-immigrants-rhetoric-criticized-el-paso-dayton-shootings/1936742001/.

Fukuyama, Francis. 1992. *The End of History and the Last Man*. New York: Free Press.

Grandin, Greg. 2007. *Empire's Workshop: Latin America, the United States, and the Rise of the New Imperialism*. New York: Owl Books.

Greenburg, Amy S. 2012. *A Wicked War: Polk, Clay, Lincoln, and the 1846 U.S. Invasion of Mexico*. New York: Alfred A. Knopf.

Haynes, Sam W., and Christopher Morris, eds. 1997. *Manifest Destiny and Empire: American Antebellum Expansionism*. College Station, TX: Texas A&M University Press.

Hayton, Bill. 2017. "The Week Donald Trump Lost the South China Sea." *Foreign Policy*, July 31.

Herring, George C. 1996. *America's Longest War: The United States and Vietnam, 1950-1975*. 3 ed. New York: McGraw-Hill.

Hoganson, Kristen L. 2000. *Fighting for American Manhood: How Gender Politics Provoked the Spanish-American and Philippine-American Wars*. New Haven, CT: Yale University Press.

Hunt, Michael H. 1987. *Ideology and U.S. Foreign Policy*. New Haven, CT: Yale University Press.

Huntington, Samuel P. 1993. "The Clash of Civilizations?" *Foreign Affairs* 72 (3): 22–49.

James, C.L.R. 1963. *Black Jacobins*. New York: Vintage Books.

Jansen, Jan C., and Jurgen Osterhammel. 2017. *Decolonization: A Short History*. Princeton, NJ: Princeton University Press.

Jardin, Ashley. 2019. *White Identity Politics*. New York: Cambridge University Press.

Joiner, Lynne. 2009. *Honorable Survivor: Mao's China, McCarthy's America, and the Persecution of John S. Service*. Annapolis: Naval Institute Press.

Kahn, E.J. 1975. *The China Hands: America's Foreign Service Officers and What Befell Them*. New York: Penguin Books.

Karp, Matthew. 2016. *This Vast Southern Empire: Slaveholders at the Helm of American Foreign Policy*. Cambridge: Harvard University Press.

Kight, Stef W., and Michal Kranz. 2020. "The Evolution of Trump's Muslim Ban." *Axios*, February 10. https://www.axios.com/trump-muslim-travel-ban-immigration-6ce8554f-05bd-467b-b3c2-ea4876f7773a.html.

Kramer, Paul A. 2006. *The Blood of Government: Race, Empire, the United States, and the Philippines*. Chapel Hill: University of North Carolina Press.

LaFeber, Walter. 1993. *Inevitable Revolutions: The United States in Central America*. 2 ed. New York: W.W. Norton.

———. 1998. *The New Empire: An Interpretation of American Expansion 1860-1898*. Ithaca, NY: Cornell University Press.

Leigh, Karen. 2020. "Trump Says He Will Stop Using the Term 'Chinese Virus." *Bloomberg.com*, March 25.

Lopez, Ian Haney. 2014. *Dog Whistle Politics: How Coded Racial Appeals Have Reinvented Racism and Wrecked the Middle Class*. New York: Oxford University Press.

Love, Eric T.L. 2004. *Race Over Empire: Racism and U.S. Imperialism, 1865–1900*. Chapel Hill: University of North Carolina Press.

Martina, Michael. 2017. "China's Xi Pledges 'Unremitting' Efforts for Middle East Peace." *Reuters*. https://www.reuters.com/article/us-china-palestinians-idUSKBN1A3176.

McCormack, John. 2016. "The Election Came Down to 77,744 Votes in Pennsylvania, Wisconsin, and Michigan." *Weekly Standard*, 10 November. https://www.washingtonexaminer.com/weekly-standard/the-election-came-down-to-77-744-votes-in-pennsylvania-wisconsin-and-michigan-updated.

Nwaubani, Ebere. 2001. *The United States and Decolonization in West Africa, 1950-1960*. Rochester, NY: University of Rochester Press.

Office of the President. 1942. "Executive Order 9066: Resulting in the Relocation of the Japanese." https://www.ourdocuments.gov/doc.php?flash=false&doc=74&page=transcript.

Okihiro, Gary Y. 2015. *American History Unbound: Asians and Pacific Islanders*. Oakland: University of California Press.

Rapoza, Kenneth. 2017. "Communist China Is Now Leader of the 'Free Trade' World." *Forbes*, 24 January.

Saletan, William. 2017. "What Trump Supporters Really Believe: The President's Racist Base, by the Numbers." Last Modified August 29. http://www.slate.com/articles/news_and_politics/politics/2017/08/trump_s_bigoted_base_by_the_numbers.html.

Scott, Joan. 1986. "Gender: A Useful Category of Historical Analysis." *The American Historical Review* 91 (5): 1053–75.

Service, John S. 1971. *The Amerasia Papers: Some Problems in the History of U.S.-China Relations*. Berkeley: Center for Chinese Studies, University of California Press.

Sides, John, and Kimberly Gross. 2013. "Stereotypes of Muslims and Support for the War on Terror." *The Journal of Politics* 75 (3): 583–98.

Takakai, Ronald. 1998. *Strangers from a Different Shore: A History of Asian Americans*. New York: Back Bay Books.

Takaki, Ronald. 1995. *Hiroshima: Why America Dropped the Bomb*. Boston: Little, Brown & Co.

Takei, Milton. 2012. "Racism and Global Warming: The Need for the Richer Countries to Make Concessions to China and India." *Race, Gender & Class* 19: 131–49.

Thomas, Hugh. 1997. *The Slave Trade: The Story of the Atlantic Slave Trade, 1440-1870*. New York: Simon & Schuster.

U.S. Government. 1882. An Act to Execute Certain Treaty Stipulations Relating to the Chinese, May 6, 1882; Enrolled Acts and Resolutions of Congress, 1789–1996. Edited by General Records of the United States Government: Record Group 11. Washington, DC: National Archives.

Woodroofe, Louise P. 2013. *Buried in the Sands of the Ogaden: The United States, the Horn of Africa, and the Demise of Détente*. Kent, OH: Kent State University Press.

7

Gender: White Women in the Age of Trump

Charlotte Chorn Dunham

Abstract There is a very close connection in American culture between racism and women's subordination. Historically, women are regarded as inferior to men as people of color are seen as inferior to Whites. Charlotte Chorn Dunham uses "controlling images" to explore the effects of cultural racism on six groups of White women. Controlling images are narratives about groups and individuals that explain "who they are" in a way that supports the gendered and racial social order as being natural and inevitable. Dunham shows how controlling images are used to stigmatize White women who seek government welfare services or otherwise violate cultural norms of Whiteness.

Keywords White women · Racism · Stigma · Controlling Images · Welfare · Social class

White women have a complicated relationship with race, which is especially evident during the age of Trump. To understand that complicated relationship is to recognize that, although the country was founded on the principles of "all men are created equal," this country has generally honored that principle for only White men. The four years of the Trump presidency have been tumultuous and painful for many, as Americans have been forced to

C. C. Dunham (✉)
Department of Sociology, Anthropology, and Social Work, Texas Tech University, Lubbock, TX, USA
e-mail: charlotte.dunham@ttu.edu

D. W. Austin and B. P. Bowser (eds.), *Impacts of Racism on White Americans In the Age of Trump*, https://doi.org/10.1007/978-3-030-75232-3_7

reflect on the meaning of core values of justice and equality because of the racial divisions exploited by Donald Trump. He set the tone when he began his campaign appealing to White Americans' racial anxiety with statements about immigrants' threat in the harshest language imaginable, calling Mexican immigrants rapists and criminals. However, "some, I assume, are good people" (Time Staff 2016). But he would not have been successful with this rhetoric if there were not already underlying racism built into the fabric of our country.

Racism and White Privilege

Racism is more than just not liking somebody because of the color of their skin. Although this is certainly part of how racism works in America, it is equally important to understand how racism is entrenched in the system. Racism is part of a system of dominance and power of one group over the other, reinforced by ideologies, informal customs, formal rules and laws, and violence that can be defined as White supremacy (Ross 2016). Under White supremacy, one does not have to practice personal discrimination or hold negative stereotypes about people of color to be racist. White supremacy happens when the rules are written so that one automatically benefits because they are White. This happens in various ways that a White person may not notice in their day-to-day life. It occurs when a White person gets pulled over by a police officer and is sent away with a warning and a smile, while a Black man will be treated with suspicion, if not worse, for the same traffic violation. A house in a White neighborhood is valued more than the same home in a Black neighborhood. Or when a Black person is denied a job for no apparent reason. Privilege means that Whites are often unaware of inequities by race because they are invisible to their day-to-day experiences. And if not invisible, Whites have the privilege of being able to ignore them. How many times have you heard a White person say that they don't understand the Black Lives Matter movement because "the police have always treated me ok"? This is like a friend saying that the government's concern about COVID-19 is exaggerated because they do not know anyone who is sick.

Intersections

When understanding White women's experiences, it is essential to take an intersectional approach to recognize the complexity of social location, based on race, class, and gender (Crenshaw 1989; Collins 1990, 2000). Racial experience is often seen as a dichotomy between Black and White (or White and other) when it is never that simple. It is misleading to expect that all White women's experiences are the same because they are White (Collins 1990, 2000). A White woman may be a single mother raising two children on minimum wage or a highly paid professional in a dual-career marriage, reflecting differences in privilege shaped by social class. Women of color and White women both are less advantaged because of the White supremacy/patriarchal system that advantages men over women. Therefore, to understand the White supremacy system, one must understand how these intersectional experiences modify it.

One way that racial inequality is reinforced is with controlling images (Collins 1990, 2000). Controlling images are narratives about groups and individuals that explain "who they are" in a way that supports the gendered and racial social order as being natural and inevitable. Controlling images are more than prejudices that affect interactions between individuals. They are social constructions about classes of people that create and justify relationships of power.

We generally think of how racism is designed to endow privilege on Whites. What is not often considered are the negative consequences that accompany that privilege. Controlling images are not the cause of racism in America but are the mechanisms by which racism is naturalized by creating narratives that justify the exploitation of other human beings. For example, slave owners created a system that exploited the free labor of slaves. They found a way to justify that exploitation as natural and inevitable by portraying Blacks as less than Whites. Controlling images are still relevant in the twenty-first century as they shape public policy (as in the case of welfare and immigration) and shape interpersonal interactions between individuals.

In this chapter, I will focus on how four controlling images are used to manage and reinforce inequities based on race and gender when considering White women's experiences in the age of Trump. Specifically, I will focus on how racism negatively affects the lives of six different groups of White women: female Trump voters, middle-class women in public spaces, welfare recipients, pregnant mothers, nursing home residents, and feminist activists.

Controlling Images

The following controlling images are relevant for understanding the role of racism in affecting White women.

Image 1: The Welfare Queen

Ronald Reagan first used the controlling image of a "welfare queen." The controlling image was based on a real person from Chicago name Linda Taylor. Taylor was a criminal who used multiple aliases to defraud the system of thousands of dollars by claiming imaginary children. She was even implicated in a murder case. Taylor was not a typical welfare recipient, but she was portrayed that way. She was a Black woman who drove a pink Cadillac and ate steak and lobster, living high off the public dole. Even though she lived very differently from most Black women on welfare, her image was used to represent *all* Black women on welfare. Her case was used to justify welfare reform and remove the alleged waste and exploitation perpetrated by irresponsible Black women (Kohler-Hausmann 2008; Briggs 2017).

Image 2: The Fragile White Woman

The controlling image of the "pure and fragile White woman" has its cultural roots in the nineteenth century "cult of true womanhood," which limited White women's participation in higher education, politics, and professional occupations and reinforced her place in the patriarchal system of White privilege. This reinforced patriarchy assures their dependence on White males for protection (Welter 1996). This controlling image portrays the middle-class White woman as morally pure and pious, whose primary role in life is wife and mother, whose proper place is in the home. This controlling image supports the notion that White women must be protected and sheltered, which is used to justify racist practices supposedly needed to protect White women (Mann 2018; Madriz 1997).

Images 3: The Dangerous Black Man

Black men have been portrayed as a threat to White male dominance and the safety of White women. Portraying Black men as dangerous rapists is part of a long-standing pattern that had a role in lynching in the past (Wells 1927, 2014) and mass incarceration in the present (Haney Lopez 2010).

Image 4: The New Mammy

As Collins explains, "the mammy image is one of an asexual woman, a surrogate mother in blackface whose historical devotion to her White family is now giving way to new expectations. Contemporary mammies should be completely committed to their jobs" (Collins 1990, 2000, 74). Research on the mammy image has shown that Black professional women feel they are expected to take a mammy role in their jobs by being pleasant and doing anything for the job (Wingfield 2007). This chapter will show how nursing home workers use the mammy image to manage stressful relationships with patients.

Four insights from the intersectional approach guide this analysis. First, women's experiences are shaped by the intersections of the positions and identities that they inhabit. This means that to have a meaningful understanding of women's experiences requires a recognition that their experiences are modified by the places they occupy, such as social class, race, ability, race, and age. Therefore, this chapter will be organized by the experiences of different groups of White women and the impact of controlling images on their well-being.

Second, White supremacy and racism are justified by controlling images. Controlling images about people of color permits Whites to justify their privilege and exercise interpersonal and structural power. For example, controlling images of Black women as lazy and irresponsible welfare queens have allowed politicians and policymakers to use that belief to justify transforming the welfare system (Briggs 2017). Controlling images of dangerous Black men justifies White women's efforts to remove innocent Black men from public space (Mann 2018) and create mass incarceration policies (Haney Lopez 2010).

Third, controlling images have been used at two levels of society. At the societal level, the narrative is used to achieve specific public policy goals or gain political advantage. At the individual level, narratives are also used in an effort for individual Whites to assert control over people of color. Furthermore, depending on controlling images is like trying to know another person from posts about them on Facebook. Controlling images portray people of color like cardboard characters without the complexity of a real person with unique life experiences, allowing them to be treated unjustly.

Fourth, White women pay the price for racism to support an inadequate welfare system and are complicit in a racist immigration policy that turns away workers that older White women depend on for later life care. This also creates barriers between White women and others they do not fully understand and prevents them from working together to achieve gender equality or even live in harmony in public places.

Effects of Controlling Images on White Women

The Women Who Voted for Trump

White women cast more votes for Donald Trump than they did for Hillary Clinton in 2016 (Pew Research Center 2018). This support was a surprise to many because Americans had a chance to vote for a woman for president of the United States who was the nominee of a major political party for the first time in history. It was predicted that several things would keep women from voting for, not the least of which was his record with women. There were his crude words on the *Access Hollywood* tapes, his sexist comments about female presidential candidates, and a woman's claim of sexual assault. Instead, 47% of White women voted for Trump, and 45% voted for Hillary Clinton (Pew Research Center 2018).[1] Preliminary exit polls from the 2020 election show the same voting pattern with about 50% of White women voting for Trump (Schmidt 2020). Some women celebrated some of his lewd statements with T-shirts and signs at rallies. For example, the New York Daily News published a picture of a woman at a Trump rally in October 2016 that said, "Hey Trump talk dirty to me!" However, one-in-three White female Trump voters expressed concern for gender equality. For example, these voters believed that there weren't enough women in political office and reported they had experienced sexist treatment (PerryUndem Institute 2017). It was clear that women who voted for Trump felt that other issues were more important and decided to overlook his sexist behavior and their concerns about gender equality (Junn 2017). The more likely problems to influence their votes were related to how White women thought about race.

Trump appealed to "the forgotten men (and women)" in society in his victory speech after the 2016 election (Gage 2016), implying that his White working-class voters who were anxious about economic insecurity were

[1] It has been widely reported that 52% of White women voted for Trump. However, this is based on exit poll data that is generally not as accurate as national polling data. Using such national data, the Pew Research Institute corrected the numbers to a smaller lead for Trump.

responsible for his victory. Indeed, White working-class Americans are experiencing a decline in life expectancy due to increases in "deaths of despair" that include death from drug overdoses, alcohol abuse, and suicide (Dossey 2018) because of economic distress. However, research examining what distinguishes White Trump voters shows that their anger is more about race than economics. Green and McElwee (2019) found that, although some of their support was based on economic concerns, the stronger predictor for Trump voters was racial anxiety. Smith and Hanley (2018) looked at White Trump voters and found that the stronger the support for Trump, the more hostility toward Blacks and immigrants. Setzler and Yanus (2018) found that most Trump voters were not exclusively working-class using the National Election Study. The strongest predictors of voting for Trump were racial resentment and sexism. Sioh (2018) claims that race was especially relevant in the 2016 election because Whites were feeling left behind and resented immigrants' presence and the increased connectedness of the United States with the world.

Trump exploited this White trauma in his campaign with his racist depictions of immigrants as criminals who unfairly used American resources and promised to build a wall to keep them out. In the 2020 presidential election, Trump encouraged dog-whistle references to dangerous threats from poor people of color aimed specifically at White women (Baker 2020). These threats appealed to the racial/ethnic anxiety of some voters.

White Women in Public Spaces

Sullivan (2014) argues racist feelings influence the "hearts and guts" of White people. Meaning that it is a mistake to ignore that White racism has a physiological component, often driven by anger and fear. A recent example of fear and anger went viral when a Black man who was bird watching in Central Park asked a White woman, Amy Cooper, to please put her dog on a leash. She was so afraid of him, she called the police and claimed he was threatening her. Her voice trembled, and it seemed that her fear was real, even if unfounded (Allen 2020). Frankenberg (1993), in her study of White women about race, found that fear of people of color was pervasive due to a geography of space that separates Blacks from Whites and encourages misunderstanding and fear. Whatever the reason, whether real fear or performance, the controlling image of the "dangerous Black man" makes her performance believable. It is also important to emphasize that women have a legitimate reason to worry about their safety in public spaces, but one has to be concerned that most of these incidents focus on Black men when we know most violent crime is perpetrated by someone of the same race (Morgan and Oudeteck 2019).

In another incident, Black Lives Matter demonstrators were in front of an expensive home (which they found out was on a private street) when the White homeowners, Mark and Patricia McCloskey of St. Louis, confronted the protestors while brandishing firearms and demanding that they leave. The protestors were not threatening, and rather than merely asking the protestors to leave, the White couple threatened them with weapons. In both situations, the Whites were middle to upper class who tried to use their White privilege to demand Blacks leave a public space (Jackman 2020). This is an example of how White women respond to the controlling image of the Black man as dangerous, which affects Black men's ability to move through public space without suspicion. And this evokes the controlling image of the fragile White woman who needs protection by the police, a White husband, or her own weapon. This focus on White women's fragility is disempowering by making women seem weak. This reinforces patriarchy because it assumes that they need White male protectors to be safe and encourages voting for someone like Trump, who promises their safety.

In some cases, White women have suffered repercussions for their behavior. When these events go viral, women have suffered consequences ranging from embarrassment to losing a job (Allen 2020) to facing weapons charges (Jackman 2020). The White woman is responding to a created reality that is harmful to herself and the community. Maintaining this narrative about Blacks ultimately creates hostile divisions in communities that distract from solving problems peacefully. Unfortunately, in the age of Trump, McCloskey's behavior was rewarded as she and her husband were invited to speak at the 2020 Republican convention (Booker 2020). They used dog-whistle narratives to allude to the danger of Blacks, which places a barrier in the way of reconciling our racial differences.

White Women in Poverty

Although a higher percentage of people of color have income below poverty, most poor people are White women. In 2019, we can see the distinct differences in poverty risks by looking at the percent in poverty by female-headed household. Black (29.5%) and Hispanic (24.6%) female-headed households have substantially higher rates of poverty compared to Whites

(16.1%) (Semega et al. 2020). However, Whites are 60.1% of the population, compared to 18.5% Hispanic and 13.4% Black (U.S. Bureau of the Census 2020). As a result, support for punitive policies because of racist images about poor Black women affects larger numbers of White women. Policymakers' reliance on the controlling image of the welfare queen shaped current welfare policy that emphasizes a transition to low-wage work rather than providing resources that would allow them to find the kind of employment to get them out of poverty altogether (Briggs 2017; Edin and Schaeffer 2015). Families headed by women with these low-wage jobs suffer from a lack of adequate childcare, sufficient food, and affordable housing. Rather than a solution to poverty, punitive welfare-to-work programs take women off the welfare rolls but increase the ranks of the permanent working poor (Edin and Schaeffer 2015).

When racist controlling images are used to achieve specific goals, it is called strategic racism. Such is the case when controlling images have been used for decades to strengthen support for policies that punish the poor. Strategic racism is related to using dog-whistle politics, where politicians know their bias and meanness would be obvious if they used openly use racist language, so they use code instead (Haney Lopez 2014). One example of this is Ronald Reagan's use of the term "welfare queen" (Briggs 2017). This is a controlling image of a Black woman who lives on welfare by manipulating and cheating the system. This image of women is used to justify the implementation of the welfare-to-work programs that have removed the social safety net for poor women (Edin and Schaeffer 2015). Although the welfare queen's controlling image assumes to be of Black women, this image implicates poor White women's deservedness.

Controlling images have been applied to poor White women by translating this racial myth into a cultural myth. Studies of the poverty of poor White women in Appalachia describe similar controlling images of "White trash" deemed lazy and promiscuous. In Hays' study of White women on welfare, a White woman liked welfare reform because it controlled other "greedy" women. This respondent created symbolic boundaries between herself and the "bad" women on welfare to avoid the stigma of being a recipient herself (Henry and Gordon 2014). Even though some Whites qualify for welfare, others who qualify refuse to use it because of what it implies about their moral character (Hayes and Hartlep 2013). When they succeeded in tying welfare recipients' image to African Americans, poor Whites perceived that being identified with Black poor people would diminish their status as Whites. Gilens (1999) found that the most highly correlated factor with these attitudes was negative attitudes toward African Americans—but no other groups of color.

The use of racist controlling images of poor Black women has negatively impacted White women. The Trump administration cut benefits even more, with proposals to cut the food stamps program (SNAP) during the COVID-19 pandemic (Picchi 2020). Images of Black women as undeserving have been applied to all poor women, with severe consequences for public assistance. The support for this controlling image of women who need help harms White women because they too must participate in this inadequate program based on a false premise about most poor women's character.

White Women and Drug Abuse

Welfare reform is not the only example of controlling images to create a punitive policy against women based on race. In the 1980s, crack cocaine emerged in poor communities of color with devastating effects. The media portrayed the typical crack users as poor, inner-city, African Americans whose drug use was responsible for crime, violence, and family decay (Logan 1999; Day and Richardson 1993). The controlling image of the bad mother/jezebel was amplified, portraying Black mothers who used crack as dangerous to their children and society. This time the panic was over stories of "crack babies," children of drug users born with severe disabilities that would affect them the rest of their lives. These children pose a huge burden on the foster care system (Day and Richardson 1993; Netherland and Hansen 2016; Logan 1999). A story published in *Rolling Stone*, October 18, 1990, described "crack babies" and their mothers in stark terms. The following excerpt suggests that babies are born unable to respond normally to caretakers, while the drug robs their mother's ability to nurture.

> Crack leaves nothing to chance. Not only does it make babies only a mother could love, it wipes out that (mother) love as well… in short, these babies lack just about everything that makes adults coo. (Hopkins 1990)

A study of *New York Times* stories about the crack epidemic written between 1985 and 2002 shows a common theme describing irresponsible crack-addicted Black mothers who neglect and mistreat their children, which justified intervention by the state. These stories often cited anecdotal evidence from medical personnel who observed the abnormal behavior of "crack babies." Anecdotes were used of Black mothers with multiple children by separate fathers to demonstrate the moral failing of Black mothers using drugs. Promotional materials for *Project Prevent* (an organization designed to provide contraception for poor Black women addicts) used these anecdotes, even though they only represented two women (Derkas 2012). The material included statements such as "we spay cats to prevent too many kittens, but we allow women to have litters of 14 children" (2012, 184). Few of these observations were backed by quality research to determine just how representative were these observations (Springer 2010).

The sad irony is that these assertions of child harm are based on myths about the effect of cocaine on the developing fetus. Research has shown few lasting effects on children born to addicted mothers, and there is more damage to infants from alcohol use and smoking in pregnancy, which is much more common (Day and Richardson 1993; Terplan et al. 2015). Still, authorities focused on illicit drug use, especially crack cocaine, as a reason for the state to intervene in a woman's pregnancy. Paltrow and Flavin (2013) reviewed legal cases from 1973 to 2005 to identify the factors in the arrests and attempts to arrest pregnant women based on concern for fetal harm. During that time, they identified 413 cases in 44 states, mostly in the South, and overrepresented by poor mothers of color reported to authorities by health care workers. However, a substantial number of these mothers were White ($N = 152$). Of the 413, 354 were charged with at least one crime, 295 of which were felonies. Authorities used existing drug and child welfare laws to target mothers ranging from child abuse, delivery of drugs to a child, and even fetal homicide in the case where a child was delivered stillborn. Most of these charges were dismissed or not upheld by appeals courts, but many poor pregnant women spent time in jail or had their children removed from the home.

The racist assumptions that poor Black women are irresponsible, immoral, and neglectful mothers that fueled the punitive and coercive policies are applied to all poor women, not just Black women. As with welfare reform, punitive policies based on controlling images originally aimed at poor Black women are used against poor White women. Bridges (2020) describes this as the "attenuation of White privilege by the exercise of White privilege." By creating drug policies designed to punish Black mothers, the same punishments apply when the drug user is a poor White mother. White privilege is not enough if a White woman is poor to protect her from the punitive laws. Poor White women are "reaping the bitter seeds of the racism that the government directed toward-and designed-for people of color" (Bridges 2020, 80).

In the 2000s, a new drug epidemic emerged, the opioid crisis, that is now the focus of public concern, but this time the media attention focuses on White users (Mendoza et al. 2018; Bridges 2020; Netherland and Hansen 2016). The response to drug use has changed from a criminal approach toward poor Black mothers' crack cocaine use to defining opioid addiction as a medical problem caused by conditions outside of the person when the typical user was White. Rather than emphasizing the drug user's moral failings, the media describes Whites as "unwitting victims" of the failings of others, including doctors and pharmaceutical companies, and unavoidable circumstances such as the need for opioids to deal with pain from a medical condition. Mendoza, Rivera and Hansen (2018) contrasted the media images of White drug users with images of people of color in the media and found that Whites were usually described as innocent victims of outside forces while people of color were portrayed as invading drug traffickers.

The paradox of drug use of middle-class pregnant mothers does not contradict beliefs in White supremacy. Changing the narrative about addiction for middle-class Whites preserves the contrast between Whites as law-abiding and people of color as the deviant other. Changing the narrative simultanously supports the system of White privilege and dominance. The White drug addict maintains their privilege of Whiteness because their addiction is defined as an exception, not the norm, and should be viewed with compassion, not as a moral failing like the poor Black mother. In a study of Staten Island media accounts (Mendoza et al. 2018), White users were portrayed as "unlikely users" who were unwitting patients because of others' mistakes, such as overprescribing physicians and overzealous and greedy pharmaceutical companies. The stories focused more on creating a humanizing and sympathetic portrait designed to evoke sympathy, not punishment, unlike Black or working-class users' stories. The contrasts between the "good" mother and the "bad" mother, the "criminal" addict, and the "unwitting" addict preserve moral distinctions between Whites and women of color that are used to justify White supremacy. Comparably, the contrasts are supported by a type of maternalism with the bad mother's image reinforcing the image of the good mother (Duggan 2016) just as the controlling image of the jezebel/bad mother supports the pure/true woman who is White.

The use of punitive and false narratives about poor Black mothers used to develop a system of social control around drug use has costs to all Americans, White and Black. With the attenuation of White privilege (Bridges 2020), White women pay considerable personal and financial costs for these laws. First, there is the enormous cost of incarceration to support punitive policies based on the belief of poor Black women users' moral failings. According to the Center for American Progress, since 1971, the United States has spent 1 trillion dollars on the war on drugs. In 2015, the federal government alone spent $9 million a day incarcerating people for drug crimes. These estimates do not include the other indirect costs of lost productivity, community crime, and family disruption, as well as additional costs of supporting the criminal justice system. Despite all the money and resources spent, research suggests incarceration is ineffective in stopping drug use (Pearl 2018).

Second, the focus on punishment has meant that the considerable resources that have been used on law enforcement and social control, while leaving little money available for investment in treatment (Bridges 2020). Third, poor White women are subject to the same punitive laws and policies as poor Black women, even though the laws were written with Black women in mind. In Paltrow and Flavin's study (2013), a considerable number of poor White women were subject to arrest for drug use while pregnant. Finally, reporting requirements compromise women's trust in health care professionals who have been required to report a mother with cocaine in her system to authorities (Paltrow and Flavin 2013; Duggan 2016). Reporting mothers' drug use to law enforcement discourages mothers from seeking prenatal care and domestic violence treatment. Such a policy is a threat to a woman and her child's health and safety (both Black and White) and is contrary to health care goals.

Elderly White Women

Racism affects White women at all stages in life, including the elderly. Because of differences in life expectancy, most nursing home residents are elderly White women (Center for Health Statistics 2019). The number of direct care workers is expected to increase by 1.1 million between 2014 and 2024 (Espinoza 2017). Most of the hands-on care workers (such as nursing aides) are people of color (59%), immigrants (24%), and females (86%). There is a severe shortage of hands-on care workers expected to worsen as the baby boomers age (Espinoza 2017). Because the industry relies on immigrants to staff these facilities, this problem has worsened because of Trump's immigration policies. Potential workers from poor countries such as El Salvador, Guatemala, and Nicaragua seeking asylum because of dangerous conditions in their home countries are detained at the border. These are the people Trump referred to as dangerous that he has built a wall to keep out. Immigrants from African countries are another source of workers who experience discrimination because of race and immigrant status (Showers 2015). These are people Trump is upset with because they "don't go back to their huts," and are from "sh**hole countries" (Watkins and Phillip 2018). Ironically, migrant health care workers are better educated than nonimmigrant workers (Zallman et al. 2019) and provide an essential labor source in health care.

Studies of nursing home workers have shown a problem with racist abuse of workers by White residents (Showers 2015; England and Dyck 2012;

Berdes and Eckert 2001; Foner 1995). In a study of three nursing homes in Indiana, Berdes and Eckert (2001) found that three-quarters of staff experienced racist incidents from residents and their family members. Unfortunately, workers experience malignant racism, which includes derogatory terms such as the "N" word, refusal of care from a person of color, or refusing to sit next to or be touched by a person of color. The authors reported that African immigrant workers have the most problems with abuse (Berdes and Elkert 2001; Wingfield 2007).

The workers themselves evoked the mammy image as a coping mechanism to deal with abusive treatment. Workers describe relationships with the residents as being like family and use language to describe family affection as a means of managing emotionally hurtful racists incidents. The workers excuse the behavior as caused by illness rather than by the person's character (Berdes and Elkhart 2007; Foner 1995). The hands-on care workers in nursing homes do not have the option to change jobs, especially the immigrant workers, and use other means of coping.

Although using the mammy image to manage stress may help in the short run, one has to question their long-term use. It prevents the real changes needed to manage worker stress, such as sufficient staffing, adequate pay, and educating residents about racism. Difficult working conditions, which include shift work, time pressures, and pressure to perform, as well as patient bullying (Kandelman et al. 2018) are related to burnout (Wolleson et al. 2019; Jetha et al; Ramirez et al. 2006) and in the most extreme cases, verbal, and physical abuse. Foner (1995), in her study of nursing home workers, found abuse is often committed by normally caring staff, who, because of the accumulation of stress, momentarily, "lose it." This research suggests that the best way to deal with abuse is to improve working conditions.

Ultimately, the stress between patients and workers is driven by issues of power and control. Elderly adults who had careers, raised children, and been involved in their communities, find themselves in positions of dependency, which they struggle to reconcile. Hostile behavior is also an outcome of the residents' stress. Because of the gendered racism, they are also inclined to lash out at the person physically closest to them, who are lower status women of color (Dunham and Cannon 2008).

The Women's March

Donald Trump was elected with the support of White women voters, yet in January of 2017, as many as 500,000 women turned out for the Women's March in Washington, DC, to oppose his policies. The day after the inauguration, women marched in the largest mass protest ever held in Washington and in over 600 cities around the world to support women's rights and protest the election of Donald Trump (Tambe 2017). However, White women who voted for Clinton have racial issues of their own.

From the beginning, there were racial tensions associated with the march. White women were the original organizers, and as often happens in feminist collaborative projects (Henry and Gordon 2014), women of color felt their concerns were ignored. These concerns threatened to derail the march until women of color were added to the planning committee. According to Presley and Presswood (2018), "social issues that ostensibly should have been a uniting factor for women eager to protest became divisive." One of the most pressing issues was the concern of unequal representation of all women at the march. In response to the criticism, the organizers changed the name of the march from Million Woman March because it was appropriating the work of Black women who named a previous march the Million Woman March. They also added three national co-chairs of color with organizing experience to make the march more inclusive, including several women of color for speakers (Presley and Presswood 2018). Although this helped make the march more inclusive, it was not satisfactory to everyone. Some saw this as indicative of White women still dominating the movement (Brewer and Dundes 2018; Cisco and Collins 2018).

Racial tensions have been part of the women's movement from the beginning, mirroring the racial inequalities in society. Early feminist leaders, although active in the abolitionist movement, still held many of the racist views of the time. White women made it clear that their intent was not to abolish slavery to give equal rights to Black men and women or to work side by side with Black feminists (hooks 1981/2015). In the 1970s, White women dominated the women's movement, putting forward issues, such as the equal opportunity for professional women, that were relevant to middle-class White women but not necessarily a priority to women of color struggling in low-wage jobs. This lack of inclusiveness has resulted in some Black women rejecting the label feminist for womanist (Walker 1983) and the development of the reproductive justice movement (Ross and Sollinger 2017).

White feminists dominated the women's movement in three ways. First, they coopted movements that were led initially by Black women. Although most of the attention to the #MeToo movement has been given to White celebrity women, the movement was started by a woman of color (Tarana Burke) to support women who were victims of sexual violence. Second, White feminists have historically erased the accomplishment of Black feminists in writing the history of the movement. Third, White feminists have claimed that we had moved on from racism and the most important challenge is sexism (Jonsson 2016). In "When Feminism is White Supremacy in Heels" Rachel Cargle said White feminists' defensiveness about acknowledging their racism makes it difficult to have honest discussions about race (Cargle 2018).

This type of racism harms White women by compromising the ability to achieve equality. Making the accomplishments of women of color invisible gives a false impression that the women's movement is solely a White women's movement, which alienates women of color. This splits the energies of activists rather than women working for the same goals. Also, when the movement is driven by White leadership based on issues important to primarily White women, they are exercising a form of White privilege that weakens the movement's moral authority. By dominating the women's movement's leadership and agenda, White feminists reproduce the imperial racism that Whites have long practiced (hooks 1981). Put another way, they are reproducing and maintaining White supremacy. This is ironic for a movement that seeks remedies for inequalities based on gender but tolerates inequalities based on race. However, the movement is changing based on the influence of women such as Patricia Hill Collins and Kimberle Crenshaw. They have encouraged the focus on intersectionality (race, class, and gender) in scholarship and research, which provides a more inclusive analytic framework.

Summary for How Racism Harms White Women

Racism Creates Fear, Anxiety, and Anger

The fear of the "other" created by controlling myths fractures relationships and communities by motivating White women to see danger and criminality in public spaces where none exists. Whether it is a Black man bird watching in Central Park or people walking down a private street, the impulse of Whites, especially White women, to want to control that space reflects fear and anxiety that is unrealistic and leads to a loss of civility. White women need to be aware of how their fear and distrust are products of racist controlling images. This is especially a problem when this anxiety is created by narratives meant to support White supremacy but keeps White people locked in their homes and threatening to their neighbors. The negative impacts of this constructed fear based on controlling images are a threat to the communities that White people live in but also expose White women to negative sanctions for making false accusations.

Racism Makes White Women Vulnerable to the Influence of Strategic Racism

In the form of dog whistle politics, strategic racism is used in a cynical attempt to achieve political and policy goals to shape a social welfare system that creates a sense that the "racial other" is undeserving of help (Haney Lopez 2014). The adoption of these controlling images in welfare reform causes the working class to reject social programs designed to help those in poverty because they have been told that those who use them are undeserving. This not only makes White women vote for people who convince them that help for those who are struggling is inappropriate and undeserved but also keeps White women from asking for help for themselves when they are struggling to avoid stigma.

Racism Supports Patriarchy

White women who voted for Trump said they were turned off by his crudeness and were troubled by allegations of sexual assault, yet they voted for him because he promised safety and security from racial others. The controlling images that create myths of fear and weakness led White women to defer their hopes of equality in return for safety and protection from the patriarchy. Without these racist controlling myths stoking fear, White women could be free to think about gender equality and vote for policies that many of them support, such as workplace fairness.

Racism Divides Groups of Women from Each Other

Since the beginning, the women's movement has been divided, with White women not giving women of color, especially Black women, equal visibility. Even in the era of Trump, the march that was supposed to reject his racism and sexism was plagued with conflict over the White organizers' role in controlling the event. Although compromise was reached, this is a problem for White women because they will never achieve their political goals if they do not find a way to work in solidarity with all women. This will require them to give up their history of moving to the front of a movement and, instead, to work as equals.

Racism Produces Bad Social Policy

In two different examples, images of Black women have led to social policies that are influenced by controlling myths. Myths about the lazy welfare queen or the Black woman who is a bad and irresponsible mother addicted to crack cocaine have led to public policies based on control and punishment rather than on working toward healthy communities.

During his term as president, Trump's leadership exacerbated the racial tensions in America. He began his campaign for president with heated rhetoric about immigrants. He ended his term encouraging the deadly attack on the U.S. Capitol, which resulted in five deaths, including that of a capital policeman. However, elections matter, and the American people did not send Trump back to the White House for a second term. For the first time in history, the United States has elected a woman of color as vice president, and the state of Georgia elected its first African American senator. The Biden Administration has promised a government that "looks like America" with appointments of women and people of color in important government positions, implicitly rejecting the power of controlling images to bias choices of our government leaders.

In this chapter, I have tried to show that controlling images about people of color create toxic conditions for people of color and White women. Policies that have used racist images of women of color have justified reducing the social safety net and have stifled the recruitment of caregivers for the elderly. Controlling images also impact interactions between White women and people of color that ultimately create more problems. Controlling images of dangerous Black men or controlling Black women caregivers generate tension and stress with negative consequences. Finally, efforts to work collaboratively among people who seek social justice are hampered by a history of racism in the women's movement, which showed in the march that was meant to resist Trump's policies. These controlling images are about power and dominance, whether in the hospital room or at the park or in a women's march, and they need to be examined and resisted for all women's well-being.

Discussion Questions

Q1: What are "controlling images," and how are they used to punish White women who "step out of line?"
Q2: How did the Trump administration use controlling images to cut food stamps and other welfare programs? How did this impact White women?
Q3: How does racism against Black women preclude many White women from accepting welfare that they are qualified to receive?
Q4: How do controlling images impact workplace relationships between White women and women of color?

References

Allen, T. 2020. "Three things Amy Cooper Did in Central Park to Damage Her Reputation and Career." *Forbes Magazine*, May 29.

Baker, P. 2020. "More Than Ever Trump Casts Himself as the Defender of White America." *New York Times*, September 6, 2020. https://www.nytimes.com/2020/09/06/us/politics/trump-race-2020-election.html.

Berdes, C., and J. Eckert. 2001. "Race Relations and Caregiving Relationships from Residents and Nurses' Aides in Three Nursing Homes." *Research on Aging* 23: 109–26.

Booker, B. 2020. "St. Louis Couple Who Waved Guns at Black Lives Matter Protestors to Speak at RNC." Accessed August 18. https://www.npr.org/sections/live-updates-protests-for-racial-justice/2020/08/18/903478960/st-louis-couple-who-brandished-guns-at-black-protesters-to-speak-at-rnc.

Brewer, S., and L. Dundes. 2018. "Concerned, Meet Terrified. Intersectional Feminism and the Women's March." *Women's Studies International Journal* 69: 49–55.

Bridges, K. 2020. "Race, Pregnancy, and the Opioid Epidemic: White Privilege and the Criminalization of Opioid Use During Pregnancy." *Harvard Law Review* 133: 770–851.

Briggs, L. 2017. *How Politics Became Reproductive Politics: From Welfare Reform to Foreclosure to Trump*. Oakland: University of California Press.

Cargle, R. 2018. "When Feminism Is White Supremacy in Heels." *Harper's Bazaar*, August 16, 2018. https://www.harpersbazaar.com/culture/politics/a22717725/what-is-toxic-white-feminism/.

Center for Health Statistics. 2019. Long Term Care Providers and Service Users in the United States, 2015–2015. February 2019, Series 3, Number 43.

Cisco, S., and J. Collins. 2018. "Editorial: The Women's March, Intersectionality and Those Left Behind." *New Horizons in Adult Education and Human Resource Development* 30: 1–21.

Collins, P. 1990/2000. *Black Feminist Thought: Knowledge, Consciousness, and the Politics of Political Empowerment*. New York: Routledge.

Crenshaw, K. 1989. "Demarginalizing the Intersection of Race and Sex: A Black Feminist Critique of Antidiscrimination Doctrine." *University of Chicago Legal Forum*: 139–168.

Day, N., and G. Richardson. 1993. "Cocaine Use and Crack Babies: Science, Media, and Miscommunication." *Neurotoxicology and Teratology* 15: 293–94.

Derkas, E. 2012. "'Don't Let Your Pregnancy Get in the Way of Your Drug Addiction': CRACK and the Ideological Construction of Addicted Women." *Social Justice* 38: 125–44.

Dossey, L. 2018. "Deaths of Despair." *Explore* 15: 1–6.

Duggan, E. 2016. "A Velvet Hammer: The Criminalization of Motherhood and The New Maternalism." *California Law Review* 104: 1299–332.

Dunham, C., and J. Cannon. 2008. "They're in Control Enough to Be in Control: Paradox of Power in Dementia Caregiving." *Journal of Aging Studies* 22: 45–53.

Edin, K., and H. Schaeffer. 2015. *$2.00 a Day: Living on Almost Nothing in America*. New York: Mariner Books.

England, K., and I. Dyck. 2012. "Migrant Workers in Home Care: Routes, Responsibilities, and Respect." *The Annals of the Association of American Geographers* 102: 1076–83.

Espinoza, R. 2017. *In Brief. Eight Signs the Shortage of Paid Caregivers Is Getting Worse*. PHINational.org.

Foner, N. 1995. *The Caregiving Dilemma: Work in an American Nursing Home*. Berkeley, CA: University of California Press.

Frankenberg, R. 1993. *Social Construction of White Women, Whiteness Race Matters*. Minneapolis: University of Minnesota Press.

Gage, B. 2016. "Who Is the Forgotten Man?" *New York Times*, 2016. https://www.nytimes.com/interactive/projects/cp/opinion/election-night-2016/could-trump-end-the-culture-wars.

Gilens, M. 1999. *Why Americans Hate Welfare: Race, Media, and the Politics of Antipoverty Policy*. Chicago: University of Chicago Press.

Green, J., and S. McElwee. 2019. "The Differential Effects of Economic Conditions and Racial Attitudes on the Election of Trump." *Perspectives on Politics* 17: 358–79.

Haney Lopez, I. 2010. "Post-Racial Racism: Racial Stratification and Mass Incarceration in the Age of Obama." *California Law Review* 98: 1023–73.

———. 2014. *Dog Whistle Politics: How Coded Racial Appeals Have Reinvented Racism and Wrecked the Middle Class*. Oxford: Oxford University Press.

Hayes, Cleveland, and Nicholas Daniel Hartlep. 2013. *Unhooking from Whiteness: The Key to Dismantling Racism in the United States*. Boston: Sense Publishers.

Henry, A., and L. Gordon (2014). *Feminism Unfinished: A Short, Surprising History of the American Women's Movement*. New York: Liveright Publishing Corporation.

hooks, b. 1981/2015. *Aint I a Woman: Black Women and Feminism*. New York: Routledge.

Hopkins, E. 1990. "Childhood's End: What Life Is Like for Crack Babies." *The Rolling Stone*. https://www.rollingstone.com/culture/culture-news/childhoods-end-what-life-is-like-for-crack-babies-188557/.

Jackman, T. 2020. "St. Louis Couple Who Aimed Guns at Protestors Charged with Felony Weapons Count." *Washington Post*, July 20, 2020. https://www.nytimes.com/2020/09/06/us/politics/trump-race-2020-election.html.

Jetha, A., L. Kernan, A. Kurowski, and ProCare Research Team. 2017. "Conceptualizing the Dynamics of Workplace Stress: A System-Based Study of Nursing Aids." *BMC Health Services Research* 17: 1–12.

Jonsson, T. 2016. "The Narrative Reproduction of White Feminist Racism." *Feminist Review* 113: 50–67.

Junn, J. 2017. "The Trump Majority: White Womanhood and the Making of Female Voters in the U.S." *Politics, Groups, and Identities* 5: 343–52.

Kandelman, N., T. Mazars, and A. Levy. 2018. "Risk Factors for Burnout among Caregivers Working in Nursing Homes." *Journal of Clinical Nursing* 27: e147–e153.

Kohler-Hausmann, J. 2008. "'The Crime of Survival': Fraud, Prosecutions, Community Surveillance, and the Original Welfare Queen." *Journal of Social History* 41: 329–54.

Logan, E. 1999. "The Wrong Race, Committing Crime, Doing Drugs and Maladjusted for Motherhood: The Nation's Fury Over Crack Babies." *Social Justice* 26: 115–38.

Madriz, E. 1997. "Images of Criminals and Victims: A Study of Women's Fear and Social Control." *Gender and Society* 11: 342–56.

Mann, J. 2018. "What's Your Emergency? White Women and the Policing of Public Space." *Feminist Studies* 44: 766–75.

Mendoza, S., A. Rivera, and H. Hansen. 2018. "Re-racialization of Addiction and the Redistribution of Blame in the White Opioid Epidemic." *Medical Anthropology Quarterly* 33: 242–62.

Morgan, R., and B. Oudeteck. 2019. Criminal Victimization, 2018. Bureau of Justice Statistics

Netherland, J., and H. Hansen. 2016. "The War On Drugs That Wasn't: Wasted Whiteness, "dirty doctors", and Race in Media Coverage of Prescription Opioid Misuse." *Cultural Medical Psychiatry* 40: 664–86.

Paltrow, L., and J. Flavin. 2013. "Arrests of and Forced Interventions on Pregnant Women in the U.S., 1973–2005: Implications for Women's Legal Status and Public Health." *Journal of Health Politics, Policy and Law* 38: 299–343.

Pearl, B. 2018. *Ending the War on DRugs: By the Numbers*. Center for American Progress.

PerryUndem Institute. 2017. *State of the Union on Gender Equality, Sexism, and Women's Rights.*
Pew Research Center. 2018. *An Examination of the 2016 Electorate Based on Validated Voters.*
Picchi, A. 2020. "Trump Administration Still Wants to Cut Food Stamps." *CBS News*, May 15, 2020. https://www.cbsnews.com/news/food-stamps-record-trump-fights-usda/.
Presley, R., and A. Presswood. 2018. *Pink, Brown, and Read All Over: Representation at the 2017 Women's March on Washington.*
Ramirez, M., J. Teresi, and D. Holmes. 2006. "Demoralization and Attitudes Toward Residents in Relation to Job Stressors and Work Resources." *Cultural Diversity in Long-Term Care* 13: 119–25.
Ross, L. 2016. "Fighting White Supremacy and White Privilege to Build a Human Rights Movement." *Understanding and Dismantling Privilege* 1: 7–11.
Ross, L., and R. Sollinger. 2017. *Reproductive Justice: An Introduction.* Oakland: University of California Press.
Schmidt, S. 2020. "The Gender Gap Was Expected to Be Historic. Instead, Women Voted Much as They Always Have." *The Washington Post*, November 6, 2020. https://www.washingtonpost.com/dc-md-va/2020/11/06/election-2020-gender-gap-women/.
Semega, J., Schreiber, M. Kollar, E. Schrider, and J. Creamer. 2020. Income and Poverty in the United States: 2019. United States Bureau of the Census.
Setzler, M., and A. Yanus. "Why Did Women Vote for Donald Trump?" *Political Science and Politics* 51: 523–527.
Showers, F. 2015. "Being Black, Foreign, and Woman: African Immigrant Identities in the United States." *Ethnic and Racial Studies* 30: 1815–30.
Sioh, M. 2018. "The Wound of Whiteness: Conceptualizing Economic Convergence as Trauma in the 2016 Presidential Election." *Geoforum* 95: 112–21.
Smith, D., and E. Hanley. 2018. "The Anger Games: Who Voted for Donald Trump in the 2016 Election, and Why?" *Critical Sociology* 44: 195–212.
Springer, K. 2010. "The Race and Class Privilege of Motherhood: The New York Times Presentations of Pregnant Drug-Using Women." *Sociological Forum* 25: 476–99.
Sullivan, S. 2014. "The Hearts and Gusts of White People: Ethics Ignorance and the Physiology of White Racism." *The Journal of Religious Ethics* 42: 591–611.
Tambe, A. 2017. "The Women's March on Washington: Words from an Organizer: An Interview with Mrinalina Chakraborty." *Feminist Studies* 43: 223–29.
Terplan, M., A. Kennedy-Hendricks, and M. Chisholm. 2015. "Prenatal Substance Abuse: Exploring Assumptions of Maternal Unfitness." *Substance Abuse Research and Treatment* 9: 1–4.
Time Staff. 2016. "Here's Donald Trump's Announcement Speech." *Time*, June 15.
U.S. Bureau of the Census. 2020. Quick facts. U.S. Bureau of the Census.
Walker, A. 1983. *In Search of Our Mothers' Gardens: Womanist Prose.* New York: Harvest Books.

Watkins, E., and A. Phillip. 2018. "Trump Decries Immigrants from "shithole countries" Coming to the United States." *CNN*, January 12, 2018. https://www.cnn.com/2018/01/11/politics/immigrants-shithole-countries-trump/index.html.

Wells, I. 1927/2014. *The Light of Truth: Writings of an Anti-Lynching Crusader*. New York: Penguin Books.

Welter, B. 1996. "Cult of True Womanhood: 1820–1860." *American Quarterly* 18: 151–74.

Wingfield, A. 2007. "The Modern Mammy and the Angry Black Man: African American Professionals' Experiences with Gendered Racism in the Workplace." *Race, Gender and Class Journal* 14: 196–212.

Wolleson, B., D. Hagemann, K. Pabst, R. Schluter, L. Bischoff, A-K Otto, C. Hold, and A. Fender. 2019. "Identifying Individual Stressors in Geriatric Nursing Staff—A Cross Sectional Study." *International Journal of Environmental Research and Public Health* 16: 3857–70.

Zallman, L., K. Finnegan, D. Himmelstein, S. Touw, and S. Woodhandler. 2019. "Care for American's Elderly and Disabled People Relies on Immigrant Labor." *Health Affairs* 38: 919–26.

8

Psychological Science: Taking White Racial Emotions Seriously—Revisiting the Costs of Racism to White Americans

Lisa B. Spanierman and D. Anthony Clark

Abstract Lisa B. Spanierman and D. Anthony Clark empirically demonstrate the psychosocial costs of racism to Whites (PCRW) theory. White Americans face negative cognitive, affective, and behavioral consequences of racism because they are in the dominant societal position. The PCRW scale measures the affective costs of racism to Whites. We learn that White empathy, guilt, and fear are meaningful aspects of White Americans' racial experiences. And when we consider White empathy, guilt, and fear in combination, five different types of White Americans emerge regarding racial orientation. These types range from blatantly racist and insensitive to those who are empathic and anti-racist.

Keywords Racism · Whites · Measures · Anti-racist · Emotions

L. B. Spanierman (✉)
Counseling and Counseling Psychology, Arizona State University, Tempe, AZ, USA
e-mail: lisa.spanierman@asu.edu

D. A. Clark
Interdisciplinary Humanities and Communication, Arizona State University, Mesa, AZ, USA
e-mail: d.anthony.clark@asu.edu

D. W. Austin and B. P. Bowser (eds.), *Impacts of Racism on White Americans In the Age of Trump*, https://doi.org/10.1007/978-3-030-75232-3_8

> [W]e all feel race because the category is produced not just 'objectively' but subjectively. Much like class and gender, race cannot come to life without being infused with emotions, thus, racialized actors feel the emotional weight of their categorical location. (Bonilla-Silva 2019, 2)

We argue in this chapter that White racial affect is key to sustaining and reproducing White supremacy (Du Bois 1903/1996). Drawing from conceptual and empirical advancements in the psychological sciences, we feature White empathy, White guilt, and White fear and their correlates. We also identify five types of White people with different racial affect profiles and bring these profiles to life with real-world examples.

White racial affect—the emotions of White supremacy—has costs for White people themselves and also for everyone else. Take the rather obvious example of Donald Trump. Trump advances views and public policies that foster fear of communities of color[1] among White people and promote the idea that Whites should ignore desires and demands for equity (Leonhardt and Philbrick 2018). Trump is acknowledged widely as expressing little to no empathy when either individuals or communities of color suffer devastating losses such as during the Covid-19 pandemic and the uprising to end police violence against Black Americans (Lee 2019). Indeed, Trump has emerged since 2015 in American cultural and political life as the embodiment of White fear and rage (Ott and Dickinson 2019). He demonstrates the kinds of fear and rage that are rooted in simultaneous reproduction and justification of willful ignorance and distorted views of reality that are dangerous and threatening to everyone.

Trump's voluminous Twitter activity offers countless examples of White fear and rage during the nation-wide reaction to the murder of George Floyd—a 46-year-old brother, cousin, and father. In particular, two tweets stand out in response to the protests erupting in the wake of Mr. Floyd's murder on Memorial Day 2020. Trump commented first on protests in Minneapolis by calling protesters "THUGS" and tweeted, "when the looting starts, the shooting starts," a line that originated in White Southern state-sanctioned violence targeting Black civil rights activists. This gives his use of the phrase an especially grim undertone rooted in disdain, disregard, and disgust for Black people. Trump recalled violent civil rights era imagery again the following day when he warned that protesters at the White House would face "vicious dogs." Trump noted "nobody came close to breaching" the

[1] When we write "people of color" and "communities of color," recognizing their rhetorical shortcomings, we use shorthand to refer to both individuals and communities of color and Indigenous Peoples.

White House fence after praising the Secret Service for handling protesters who "got too frisky or out of line" by coming "down on them, hard [so that they] didn't know what hit them." "If they had, they would have been greeted with the most vicious dogs, and most ominous weapons." Rooted in White fear of the racialized other, Trump and his most vocal supporters seem to experience pleasure in the imagery and reality of White rage and racist domination.

Trump's racial affect and its cost to the communities of color and multiracial coalitions they target is obvious. Less plain, perhaps, are the costs of White fear and rage to Trump himself and his followers. White racial fear is linked to a false sense of reality—one in which White people erroneously see themselves as superior to everyone else. White fear also is isolating, as it deprives Whites who believe they are superior from experiencing mutually meaningful intercultural and multiracial relationships. White fear is dehumanizing to people and communities of color, but also to White people themselves.

Of course, not all White people are Trump followers. Fueled by negative stereotypes in media, for instance, most White people experience some degree of fear of people of color—especially of Black and Brown men (e.g., Fanon 1952/2008; Smith et al. 2016). In the current moment, this White fear is most evident in police brutality that targets and dehumanizes people and communities of color. A deep-seated White fear surely plays a role in a police officer's decision to use levels of force that too often end in unjustified homicides. While some law enforcement officers certainly express overt racist attitudes rooted in White fear and rage that are akin to Trump's racist utterances, others hold more implicit biases for which they themselves may be unaware (e.g., Goff et al. 2014).

As we seek to understand the recent solidarity of many White people marching in support of racial justice alongside Black and other communities of color, White racial affect again comes to mind. In particular, these White demonstrators experience and express some degree of racial empathy—a seemingly heartfelt understanding and compassion for Black people's pain and anger in the face of systemic racial injustice. While Trump stoked White fear and rage in the wake of Mr. Floyd's murder, corporations, public figures, and countless White people joined Black Lives Matter to say, in the words of then newly minted WarnerMedia CEO Jason Kilar, "I want to be very clear in saying that our responsibility to the Black community is that we truly listen, that we seek understanding, and that we do these things *with as much empathy as possible* [our emphasis]." Also, some White protesters may experience productive forms of White guilt. Drawing from Baumeister et al.'s (1994) classic review of the interpersonal nature of guilt, it is likely that White guilt could function to repair relationships with individuals and communities of color and propel action toward realizing racial justice.

In what follows we address research on the affective costs of racism to White people since 2004. Scholars in recent years have drawn from empirical and conceptual literature to outline a taxonomy of the emotions of racism and antiracism for White people (Spanierman and Cabrera 2015). Others have focused empirically on specific White racial emotions such as guilt and shame (e.g., Grzanka et al. 2020) and fear (e.g., Siegel and Carter 2014). Only psychosocial costs of racism to Whites theory and research thus far has focused specifically on understanding White empathy, guilt, and fear collectively in the context of a racialized social system (Bonilla-Silva 1997).

Psychosocial Costs of Racism to Whites: A Focus on Racial Affect

Although social scientists properly acknowledge that White Americans receive a host of unearned privileges, it may be less apparent that individual Whites also experience *negative* consequences from living in such a system. Spanierman and Heppner (2004) theorized an approach to understanding these negative consequences—deemed costs of racism to individual Whites. The term "costs" here does not refer to any sort of "reverse racism" that allegedly targets Whites; White people do not suffer the debilitating material effects of White supremacy. Rather, "costs" refers to the negative cognitive, affective, and behavioral consequences experienced by Whites as dominant group members in a White supremacist system. Recognizing that racism affects *everyone* in detrimental ways is important, while the costs of racism to Whites are in no way comparable to the costs of racism that people of color face on a daily basis and across generations. Highlighting the costs of racism to Whites has the potential to enhance the relevance of messages about racial justice for Whites who otherwise may think that racism affects only people of color. Below, we first address briefly the scholarly literature in psychology and education that informs costs of racism to Whites theory. Next, we describe a scale designed to measure affective (emotional) costs of racism to Whites and detail empirical findings from the scale.[2] Finally, we draw from research on racial affect profiles, or types, to speculate about how utterances and behaviors since Mr. Floyd's murder in Minneapolis evidence the five types of White Americans.

[2] We also explain a statistical clustering approach that uses the costs of racism scale scores to identify five types of White people.

Three early conceptualizations of the costs of oppression to dominant-group members published between 1981 and 2001 highlighted the broad ways in which systems of domination (e.g., racism, sexism, and heterosexism) affected Whites, men, and straight individuals and are key to the psychosocial costs of racism to Whites theory (see Bowser and Hunt 1981, Hunt and Bowser 1996; Goodman 2001, 2011; Kivel 1996, 2017). In the first edition of *Impacts of Racism on White Americans*, Bowser and Hunt (1981) introduced scholarship on the various negative consequences that Whites experience from sitting atop a racial hierarchy. For example, Karp (1981) described emotional effects such as isolation, guilt, and shame. Dennis (1981) observed that racism promotes a "double social and psychological consciousness" among Whites where they suffer from a split between deed and creed (e.g., perpetuating racism while believing in freedom and democracy for all). Similarly, Terry (1981) observed a condition that he termed "White inauthenticity" wherein Whites are not true to themselves or the world (e.g., belief in the myth of meritocracy). During the closing decade of the twentieth century, antiracist educators Paul Kivel (1996) and Diane Goodman (2001) further explicated costs of oppression to dominant group members. Kivel (1996) coined the phrase "costs of racism to Whites" and created a checklist of these costs: false sense of superiority, inaccurate view of history, distorted sense of danger, and loss of moral integrity. Similarly, Goodman (2001) identified broad categories of costs of oppression to people from privileged groups: psychological, intellectual, material, social, and spiritual.

Extending this earlier work, Spanierman and Heppner (2004) used the established psychological tripartite framework (i.e., cognitive, affective, and behavioral) to delineate the psychosocial costs of racism to White people. *Cognitive costs* refer to distortions in thinking about oneself (e.g., believing that all achievements are merit-based), misunderstandings and misrepresentations of people of color (e.g., relying on stereotypes), or inaccurate views of reality (e.g., denying the existence of racism). *Affective costs* signify emotional consequences of racism, such as sadness or anger about the existence of racism, guilt about one's racial privilege, and irrational fear of people of color. *Behavioral costs* of racism involve limiting one's actions, such as censoring oneself in interracial settings and having close friendships only with White people. We hope it is clear that costs of racism to Whites are conceptualized differently than the ways in which people of color experience the abhorrent effects of racism on a daily basis and that people and communities of color pay an additional price for the costs of racism that White people experience.

Measuring Costs of Racism to Whites

Psychological researchers develop scales to operationalize constructs, test theory, explore associations with other constructs, and predict attitudes and behaviors. Thus, Spanierman and Heppner (2004) introduced the Psychosocial Costs of Racism to Whites (PCRW) scale to quantify and measure negative consequences of racism to Whites. Scale development is onerous and data-driven. Data from more than 700 observations resulted in a measure focused on affective, or emotional, costs of racism. Because emotional responses are particularly salient among Whites with regard to racial issues, and few measures of racial affect were available, Spanierman and Heppner (2004) shifted their focus from the tripartite model (i.e., cognitive, affective, and behavioral) to understanding affective costs of racism to Whites.

The 16-item, self-report instrument comprises three subscales: White empathy, White guilt, and White fear. *White empathy* refers to anger and sadness about the existence of racism. The empathy subscale consists of six items including, "I am angry that racism exists," "It disturbs me when people express racist views," and "I become sad when I think about racial injustice." *White guilt* involves experiencing remorse, shame, and responsibility about one's privileged position in a racialized social system. The guilt subscale consists of five items including, "Sometimes I feel guilty about being White," "I never feel ashamed about being White (reverse scored)," and "Being White makes me feel personally responsible for racism." White fear pertains to irrational mistrust, anxiety, and avoidance of people of color. *White fear* comprises five items including, "I am distrustful of people of other races," "I am fearful that racial minority populations are rapidly increasing in the U.S., and my group will no longer be the numerical majority," and "I have very few friends of other races." Respondents select a score from 1 (*strongly disagree*) to 6 (*strongly agree*), and higher scores indicate greater experiences with White empathy, guilt, or fear. Across studies, we have learned that White empathy, guilt, and fear are related but distinct constructs. Because the three scales operate differently, they cannot be combined for a total scale score.

Evidence across multiple studies has supported the PCRW scale's reliability and validity.[3] Next, we describe empirical findings from these studies

[3] The scale garnered initial psychometric support from a large, college-student sample (Spanierman and Heppner 2004), and subsequent investigations demonstrated adequate reliability and validity primarily among college undergraduate and graduate students (Case 2007; Chao et al. 2015; Garriott et al. 2016; Mekawi et al. 2016; Paone et al. 2015; Sifford et al. 2009; Spanierman et al. 2006; Spanierman et al. 2008; Todd et al. 2011; Turner 2011). Additional validation studies examined the psychometric properties of the scale among a geographically dispersed sample of employed adults

detailing what we know about White empathy, White guilt, and White fear. We also describe five types of White Americans based on their racial affect scale scores.

Affective Costs of Racism: What Do We Know?

In more than two dozen quantitative investigations since the development of the PCRW scale, researchers found that demographic factors and various racial attitudes were associated with White empathy, White guilt, and White fear. Across samples of students and employed adults, for example, women scored significantly higher than men on White empathy (Chao et al. 2015; Paone et al. 2015; Sifford et al. 2009; Spanierman and Heppner 2004; Spanierman et al. 2006, 2008, 2009, 2012). In at least two studies, women also scored higher than men on White guilt (Sifford et al. 2009; Spanierman and Heppner 2004). Higher education levels and greater multicultural education have been associated with higher White empathy and lower White fear (Sifford et al. 2009; Spanierman and Heppner 2004). Additionally, greater levels of exposure to and friendships with people of color have been linked to higher levels of White empathy (Chao et al. 2015) and lower levels of White fear (Spanierman and Heppner 2004; Turner 2011). The quality of intergroup contact also predicted White empathy (Merrilees et al. 2018). Self-described Republican voters tend to score lower on White empathy and higher on White fear than self-described Democrats and Independent voters (Sifford et al. 2009).

White empathy and guilt tend to be associated with positive racial attitudes, whereas White fear has been linked to negative racial attitudes. For example, White empathy and White guilt both have been linked to greater awareness of racism and White privilege, cultural sensitivity, and openness to diversity (Black 2018; Chao et al. 2015; Pinterits et al. 2009; Poteat and Spanierman 2008; Spanierman and Heppner 2004). On the other hand, White fear has been associated with lower levels of racial awareness, cultural sensitivity, openness to diversity, and ethnocultural empathy (Black 2018; Poteat and Spanierman 2008; Spanierman and Heppner 2004).

Psychological and marketing researchers also have shown how racial affect is tied to behaviors or behavioral intentions. For example, White empathy has been linked to how White participants respond to a restaurant server's potentially discriminatory actions toward Black patrons (Baker and Meyer

(Poteat and Spanierman 2008), Ashkenazi Jews (Berk 2015), and sexual minority men (Kleiman et al. 2015).

2011). Using an experimental design, Baker and Meyer found that White participants high in White empathy, but not those low in White empathy, reacted similarly to Black participants in assessing the severity and discriminatory nature of the event. In another study, White guilt was linked to disordered eating (i.e., bulimia; Lydecker et al. 2014). White empathy and White guilt both have been linked to greater willingness to confront White privilege (Pinterits et al. 2009) and to cultural competence among mental health trainees (Spanierman et al. 2008). On the other hand, higher levels of White fear have been associated with a lower threshold for firing a weapon at a Black person relative to shooting White and East Asian individuals (Mekawi et al. 2016). Thus, racial affect seems to have a wide range of consequences for White people's behavior across various contexts.

Burgeoning investigation has documented the impact of educational interventions on these three dimensions of White racial affect. Generally, studies found that multicultural instruction (i.e., courses, workshops, and other educational interventions) increased White empathy and White guilt. In a race-based course among master's-level counseling students, for instance, Paone and colleagues (2015) found that White guilt increased from the beginning to end of the semester; this increase in White guilt accompanied other positive racial attitude changes, such as greater awareness of White privilege. Using randomized experimental designs, two independent research teams found significant changes in racial affect after college students viewed the 19-minute ABC Primetime Live Special *True Colors* on racial discrimination and privilege (Garriott et al. 2016; Soble et al. 2011). Soble and colleagues (2011) found that White empathy and White guilt increased after the educational video, but there was no change in White fear. Garriott and colleagues (2016) found similar results for White guilt. Case (2007), too, found that White guilt increased from the beginning to end of a required undergraduate diversity course. Case also found White fear increased at the end of the term and speculated that students may have been more honest with themselves at the end of the semester. In short, experimental designs generally provide support for the notion that educational interventions can facilitate desirable changes in White students' racial affect.

Racial Affect: The Five Types of White Americans

Because the PCRW subscales function in different directions (i.e., White empathy and guilt operate differently than White fear), a total racial affect score cannot be calculated. Thus, to understand the affective costs of racism to Whites in combination, Spanierman and colleagues (2006) used cluster analysis to identify groups of similar White Americans based on their racial affect scores.[4] The analysis identified five distinct racial affect types of White people that represent different constellations of White empathy, White guilt, and White fear. See Table 8.1 for a summary and examples of the five racial affect types. Generally, the *insensitive and afraid* and the *oblivious* types are blatantly racist, whereas the *empathic and unaccountable* and *fearful guilt* types show some racial awareness and sensitivity. The *antiracist* type is the only type that evidences what could be understood as critical racial consciousness. Below, we define and provide research findings on the five racial affect types ranging from the least to most racially aware and sensitive.

Insensitive and Afraid Type

Whites who evidence the *insensitive and afraid* type are likely the most racially uninformed among all White people. Research participants who have scored in this type exhibit the lowest levels of White empathy and White guilt, and the highest level of White fear, among all five types. This type, along with the *oblivious* type described below, exhibits blatant racist attitudes and expresses a high level of irrational fear of people of color. For example, *insensitive and afraid* types might declare that Black people as a whole are dangerous criminals. In focus group discussions, White people who score in this type ground their fear in actual or even made up personal anecdotes such as, "I know people who have been attacked by Black people." Because they have no awareness of institutional racism, and deny the existence of racism generally, they blame people of color for racial inequities. In fact, they are likely to make claims of reverse racism in response to public policies that seek to close racial gaps in wealth, education, health care, and other markers of well being. They might say and believe that a White person lost out on a job opportunity because a company needed to "meet its quota." At least one study showed a higher proportion of men than women in this type (Spanierman et al. 2012). Not surprisingly, among all five types, *insensitive and afraid* individuals report the lowest levels of multicultural education, cultural sensitivity, racial awareness, support for affirmative action, and exposure to people of color. This type represents the highest proportion of self-identified Republicans and the lowest proportion of Democrats (Spanierman et al. 2006), and we contend that the most ardent Trump supporters likely epitomize this type.

[4] Cluster analysis is a statistical approach that classifies data into groups to maximize similarity within a group and dissimilarity across groups.

Table 8.1 Five racial affect types of White people

Type	Characterization from the research	Real-world examples
Insensitive and afraid	• Highest fear, lowest empathy, lowest guilt • Greater proportion of men than women • Lowest levels of multicultural education and cultural sensitivity • Mostly White friends and little exposure to people of color • Greatest proportion of Republicans and smallest proportion of Democrats • Highest levels of racial colorblindness	• Fox News host Tucker Carlson decrying a third-degree murder charge against a White Minneapolis police officer who crushed the life out of a Black man by asking viewers "Why doesn't anybody stand up for the rest of us, for civilization?" • "There's a woman pushing a shopping cart in front of my house," a White man from a wealthy Oakland suburb told a 17-year veteran 911 dispatcher. "I'm sorry, I don't understand what you're reporting," the dispatcher responded. "She's Black," said the caller. The exchange continued: "Sir, I'm still not seeing the problem. Is she being loud? Is the noise of the cart disturbing your peace?" His tone got harsher. "Where do you live?" I was so startled by the question that I answered it. "Oakland," I said. "You wouldn't understand, then. This isn't Oakland. We don't have people like her in this neighborhood. Just send someone out to get rid of her. I'm not talking to you anymore."

Type	Characterization from the research	Real-world examples
Oblivious (unempathic and unaware)	• Moderate fear, low empathy, low guilt • Greater proportion of men than women • Low levels of multicultural education and cultural sensitivity • Mostly White friends; exposure to people of color was mixed across the sample • High levels of racial colorblindness • One of the most prevalent types	• While he subsequently apologized, when first asked about his thoughts were on the subject of players potentially kneeling again during the national anthem in the wake of the civil unrest that arose after four Minneapolis police officers murdered George Floyd, NFL quarterback Drew Brees told an interviewer: "I will never agree with anybody disrespecting the flag of the United States of America or our country. [I]s everything right with our country right now? No, it's not. We still have a long way to go, but I think what you do by standing there and showing respect to the flag with your hand over your heart, is it shows unity. It shows that we are all in this together, we can all do better and that we are all part of the solution." • Former Franklin Templeton investment-fund manager calling the police on a New York City Audubon Society board member and bird-watcher in Central Park who asked her to leash her dog to report "an African American man threatening my life" before later telling cable news "I'm not a racist" and "I did not mean to harm that man in any way." • Identifying "as somebody who grew up in a very diverse background as a young boy in the projects" when responding to a question from a CNN town hall audience member about a racial profiling incident that occurred at a Philadelphia Starbucks location, former Starbucks CEO Howard Schultz tells viewers he "honestly [doesn't] see color now."

(continued)

Table 8.1 (continued)

Type	Characterization from the research	Real-world examples
Empathic but unaccountable	• High empathy, low guilt, low fear • Balanced proportion of women and men • Some multicultural education • Friendships with people of color • Moderate support for affirmative action • Some understanding of racial issues, but not White privilege • One of the most prevalent types	• "If you have diabetes, obesity, hypertension – and diabetes and hypertension are clearly risk factors for problems from COVID – then African Americans are going to have…an overrepresentation [in death from COVID-positive cases]," Louisiana Republican Senator Bill Cassidy told NPR's Morning Edition, blaming Black people for having underlying medical conditions but, crucially, not explaining why. • A clearly distraught Ellen Degeneres telling viewers that the attention brought to the systemic nature of police brutality toward Black people was really about her commitments to fairness and justice for all: "I am so sad, and I am so angry, and I know I'm not going to say the right thing…. But I have a platform and I have a voice, and I have always stood for equality. I have always wanted to be the voice for people who felt like they didn't have a voice, because I know what that feels like…. I really don't know what to say other than this has gone on way, way, way, way too long. People have gotten away with murder, and that's what's happening. So, we've got to see fairness and we've got to see justice for all, because right now, this is not a fair world. Not at all. And if you don't understand this, then you've never felt like you weren't heard, or you weren't equal. But if you've ever felt that, magnify it and see what's happening. And let's send a whole bunch of love out there and try to find some peace and some communication in this."

Type	Characterization from the research	Real-world examples
Fearful guilt	• High fear, high guilt, moderate empathy • Greater proportion of women than men • Some multicultural education • Mostly White friends • Moderate support for affirmative action • Higher proportion of Democrats than Republicans • Low levels of racial colorblindness; some awareness of and remorse about White privilege	• Seattle Mayor Jenny Durkan tweeting her characterization of who was looting during the civil unrest sweeping the nation after police officers murdered George Floyd in Minneapolis while, at the same time, ignoring her complicity as mayor and a White woman: "I want to acknowledge that much of the violence and destruction, both here in Seattle and across the country, has been instigated and perpetuated by White men" and "These individuals experience the height of privilege and are co-opting peaceful demonstrations that were organized by and meant to center people of color, particularly Black Americans." • Cincinnati Reds first baseman Joey Votto describing his "awakening" in a guest column for the *Cincinnati Enquirer*: "Everything inside of me wants things to go back to normal. I don't want to protest, raise my voice, or challenge someone. I don't want to have heated arguments, break up friendships, or challenge previous norms.... But I hear you now, and so that desire for normalcy is a privilege by which I can no longer abide. That privilege kept me from understanding the 'why' behind Colin Kaepernick's decision to kneel during the national anthem. That privilege allowed me to ignore my Black teammates' grievances about their experiences with law enforcement, being profiled, and discriminated against. And that privilege has made me complicit in the death of George Floyd, as well as the many other injustices that Blacks experience in the U.S. and my native Canada."

(continued)

Table 8.1 (continued)

Type	Characterization from the research	Real-world examples
Antiracist	• High empathy, highest guilt, low fear • Greater proportion of women than men • Highest levels of multicultural education • People of color as close friends • Support for affirmative action • Greatest proportion of Democrats and smallest proportion of Republicans • Lowest levels of racial colorblindness • Keenly aware of and willing to confront White privilege	• Governor Tim Walz tweeting during the first week of police brutality protests that swept the nation in May and June 2020: "Minnesota consistently ranks highly for our public schools, innovation and opportunity, and happiness – if you're White. If you're not, the opposite is true. Systemic racism must be addressed if we are to secure justice, peace, and order for all Minnesotans" • "The President just used a Bible and one of the churches of my diocese as a backdrop for a message antithetical to the teachings of Jesus and everything that our church stands for," Episcopal Diocese of Washington Bishop Mariann Edgar Budde tweeted in response to Donald Trump's widely reported photo opportunity taken after he ordered that peaceful protesters be removed from Lafayette Park so he could walk from the White House to the St. John's Episcopal Church. "I am outraged. The President did not...acknowledge the agony and sacred worth of people of color in our nation who rightfully demand an end to 400 years of systemic racism and White supremacy in our country"

Oblivious (Unempathic and Unaware) Type

Closely aligned with the *insensitive and afraid* type, *oblivious* individuals also are uninformed about and deny that racism plays a role in structuring social relations and institutions. Low levels of White empathy and guilt with moderate levels of White fear characterize this type. The key factor that differentiates the *oblivious* type from the *insensitive and afraid* type is that White fear is moderate instead of high among *oblivious* persons. Additionally, individuals who score in *the oblivious* type report varying levels of exposure to people of color, whereas *insensitive and afraid* individuals seem to live in a White bubble. The *oblivious* type reflects one of the highest proportion of men from prior research samples. Persons in the *oblivious* type are unaware of White racial privilege, report little multicultural education, and experience few cross-racial friendships (Spanierman et al. 2006). Thus, like the *insensitive and afraid*, the *oblivious* type embraces color-blind racial ideology and as such distorts, denies, and minimizes race and racism in US society (see Bonilla-Silva 2017; Neville et al. 2013). In a focus group study among White university students, for example, participants in the *oblivious* type characterized as problematic protests calling attention to racial and social justice issues on campus (Kordesh et al. 2013). Additionally, an *oblivious* participant in the study was angry and offended that we even asked him if he felt guilty about being White. He explained that he viewed himself as "American," not as White, and believed the world would be a better place if others followed suit. Interestingly, at the same time he denied he had a race, he also claimed that he had been a victim of reverse racism.

Empathic but Unaccountable Type

Whites who are *empathic but unaccountable* usually externalize their disdain for racism, deny their own racist beliefs and actions, and absolve themselves of any responsibility for perpetuating a racist system. This type exhibits high White empathy, without the accompanying White guilt described below in the *antiracist* type. In contrast to the *insensitive and afraid* and *oblivious* types, *empathic but unaccountable* individuals report racial diversity in their friendship group and a rudimentary knowledge and awareness of race and racism. However, *empathic but unaccountable* individuals lack awareness of their White privilege in a racialized social system. While persons in this type tend to recognize individual acts of racism (e.g., "My roommate used the N-word to describe someone on our floor"), they are unlikely to acknowledge institutional racism. In a focus group study, *empathic but unaccountable* university students expressed subtle forms of racism without realizing it. They used racially-coded language to describe "shady characters" in "bad neighborhoods," for instance, linking criminal activity to geographic location and individual actors without attribution to race (Kordesh et al. 2013).

Fearful Guilt Type

The *fearful guilt* type expresses paralyzing anxiety in response to learning about or attempting to articulate their complicity in a racist system. This type scores high in White guilt and White fear and low in White empathy. In prior research, there has been a greater representation of women than men in this type. Notably, persons in the *fearful guilt* type report fewer cross-racial friendships than the *empathic but unaccountable* type but tend to expect their friends of color to teach them about race and racism. *Fearful guilt* is associated with at least some multicultural education and low levels of racial colorblindness with awareness of and remorse about their racial privilege.

In *Sister Outsider*, Audre Lorde (2012) described White guilt in a way that helps differentiate the *fearful guilt* and *antiracist* type below. She wrote, "Guilt is not a response to anger; it is a response to one's own actions or lack of action. If it leads to change then it can be useful, since it is then no longer guilt but the beginning of knowledge. Yet all too often, guilt is just another name for impotence, for defensiveness destructive of communication; it becomes a device to protect ignorance and the continuation of things the way they are, the ultimate protection for changelessness" (p. 130). Here, Lorde's use of guilt leading to change exemplifies the *antiracist* type. More common, however, is guilt that re-inscribes the status quo. We argue the impotent sort of guilt is evident among individuals in the *fearful guilt* type.

Antiracist Type

Among all five types, the *antiracist type* is the only one that understands institutional racism and takes action to disrupt racism. Research participants in the *antiracist* type report the highest levels of White empathy and White guilt and the lowest levels of White fear among all five types. Women and Democrats tend to be overrepresented in the *antiracist* type. In one study, women had 2.18 times greater odds than men of being *antiracist* than *insensitive and afraid* (Spanierman et al. 2009). Not surprisingly, individuals in the *antiracist* type report the highest levels of racial diversity among their friendship groups. Researchers also have linked the antiracist type with higher levels of religious liberalism (in contrast to religious conservatism; McConnell and Todd 2015).

Undergraduate focus group participants who scored in the *antiracist* type expressed a critical understanding of racial issues and White privilege (Kordesh et al. 2013). Unprompted, they engaged in a nuanced discussion of various campus racial issues. They also recognized coded and subtle racist practices in the local community, such as nightclubs banning persons wearing hoodies from entering. They not only acknowledged White privilege, but they discussed how they personally benefit from White privilege and expressed remorse about and willingness to confront such privilege. Their views, they told researchers, contrasted sharply with other White people in their lives. They were embarrassed by Whites who expressed racist views and often times had difficulties in their relationships with White people (especially family and friends from their hometown). They were angry about racism on campus, expressed frustration about lack of student involvement in social justice, and joined in solidarity with groups to fight oppression on campus. These *antiracist* individuals completed many more diversity-related courses than their non-*antiracist* type counterparts, and they wanted the university to require additional diversity courses with a social justice focus for all students.

Racial Affect Types: Can Whites Change?

Because evidence suggests that particular constellations of White racial emotions have implications for White people's behaviors, understanding what might increase White empathy and guilt, while decreasing White fear, has implications for developing White racial justice allies. Research indicates that an individual's racial affect type is not static. In a study among college students, for instance, Spanierman and colleagues (2009) explored patterns of change during the first year of college. First, they examined possible predictors of White students' racial affect type at the start of college. They found that openness to and appreciation of diversity at entrance predicted types that were high in White empathy (i.e., *empathic but unaccountable* and *antiracist*), whereas awareness of racial privilege predicted types high in White guilt (i.e., *fearful guilt* and *antiracist*). Additionally, they examined whether membership in a type at entrance predicted any diversity-related outcomes throughout the first year of college. They found that *empathic but unaccountable* and *antiracist* types predicted engaging in cross-racial friendships, such that high White empathy types at the beginning of the year predicted greater engagement in cross-racial friendships at the end of the year.

In addition to exploring predictors and outcomes of racial affect type, Spanierman and colleagues (2009) also explored whether White undergraduates changed racial affect type during the course of their first year on campus. The types were relatively stable and changes, for the most part, were in desired directions. For example, persons in the *insensitive and afraid* type were more likely to move to *empathic but unaccountable* during the course of the year; and, those in *empathic but unaccountable* were more likely to move to *antiracist*. Interestingly, there were some undesired changes such that those in the *oblivious* type were more likely to move to *insensitive and afraid type* than other types. Perhaps this finding can be explained by perceived threat or competition from students of color. It is also possible that dissonance-inducing events such as witnessing racism or learning about their participation in a racist system could have triggered a retreat to strong White superiority beliefs (see Helms 1995). More research is needed to understand what social and academic experiences contribute to both desired and undesired changes in racial affect types, which has the potential to inform educational interventions.

Conclusion

While we have far more questions about this matter than we have answers, research to date has provided important tools and key insights. Psychosocial costs of racism to Whites theory, for instance, indicates that even as people of color bear the burdens of White racial affect, White people themselves do face negative cognitive, affective, and behavioral consequences of racism as a result of being in the dominant position in a racialized social system. The PCRW scale measures the affective costs of racism to Whites. From research using this measure, we have learned that White empathy, White guilt, and White fear are meaningful aspects of White people's racial experiences. And when we consider White empathy, guilt, and fear in combination, we know that five different types of White Americans emerge. These types range from blatantly racist and insensitive to empathic without any personal accountability to a more informed empathy and critical analysis with accountability and willingness to act. Thus, White racial affect is likely both a curse and promise in the Age of Trump. As curse, it is foundational to reproducing material inequities, as happens when oblivious types report Black people to the police for innocuous activity, thereby potentially subjecting them to greater law enforcement surveillance and even police violence. As promise, White racial affect offers the possibility that White people can become part of the solution rather than remaining complicit with the problem of racism that they create or support. This has been indicated by the unprecedented number of antiracist Whites who have joined the protests against police violence in the wake of the deaths of George Floyd, Breonna Taylor, and countless others.

References

Baker, T. L., and T. Meyer. 2011. "White Response to Potentially Discriminatory Actions in a Services Setting." *Psychology & Marketing* 28 (2): 188–204.

Baumeister, R. F., A. M. Stillwell, and T. F. Heatherton. 1994. "Guilt: An Interpersonal Approach." *Psychological Bulletin* 115 (2): 243–67.

Berk, E. T. 2015. "Construct Validation of the Psychosocial Costs of Racism to Whites Scale for Ashkenazic Jews in the United States." Doctoral dissertation, Seton Hall University. ProQuest Dissertations Publishing.

Black, W. W. 2018. "An Examination of Relations Among Fear, Guilt, Self-Compassion, and Multicultural Attitudes in White Adults." Doctoral dissertation, University of Kentucky. ProQuest Dissertations Publishing.

Bonilla-Silva, E. 1997. "Rethinking Racism: Toward a Structural Interpretation." *American Sociological Review* 62 (3): 465–80.

Bonilla-Silva, E. 2017. *Racism without Racists: Color-blind Racism and the Persistence of Racial Inequality in the United States*. Rowman & Littlefield Publishers.

Bonilla-Silva, E. 2019. "Feeling Race: Theorizing the Racial Economy of Emotions." *American Sociological Review* 84 (1): 1–25.

Bowser, B. P., and R. G. Hunt. 1981. *Impacts of Racism on White Americans*. Thousand Oaks, CA: Sage.

Case, K. A. 2007. "Raising White Privilege Awareness and Reducing Racial Prejudice: Assessing Diversity Course Effectiveness." *Teaching of Psychology* 34 (4): 231–35.

Chao, R. C.-L., M. Wei, L. B. Spanierman, J. Longo, and D. Northart. 2015. "White Racial Identity Attitudes and White Empathy: The Moderation Effect of Openness to Diversity." *The Counseling Psychologist* 43 (1): 94–20.

Dennis, R. W. 1981. "Socialization and Racism: The White Experience." In *Impacts of Racism on White Americans*, edited by B. J. Bowser and R. G. Hunt, 71–85. Beverly Hills, CA: Sage.

Du Bois, W. E. B. 1996. *The Souls of Black Folk*. New York: Penguin Books (Original Work Published 1903).

Fanon, F. 2008. *Black Skin, White Masks*. New York: Grove Press (Original Work Published 1952).

Garriott, P. O., S. Reiter, and J. Brownfield. 2016. "Testing the Efficacy of Brief Multicultural Education Interventions in White College Students." *Journal of Diversity in Higher Education* 9 (2): 158–69.

Goff, P. A., M. C. Jackson, B. A. L. Di Leone, C. M. Culotta, and N. A. DiTomasso. 2014. "The Essence of Innocence: Consequences of Dehumanizing Black Children." *Journal of Personality and Social Psychology* 106 (4): 526–45.

Goodman, D. J. 2001. *Promoting Diversity and Social Justice: Educating People from Privileged Groups.* Thousand Oaks, CA: Sage.

Goodman, D. J. 2011. *Promoting Diversity and Social Justice: Educating People from Privileged Groups.* 2nd ed. Thousand Oaks, CA: Sage.

Grzanka, P. R., K. A. Frantell, and R. E. Fassinger. 2020. "The White Racial Affect Scale (WRAS): A Measure of White Guilt, Shame, and Negation." *The Counseling Psychologist* 48 (1): 47–77.

Helms, J. E. 1995. "An Update of Helms's White and People of Color Racial Identity Models." In *Handbook of Multicultural Counseling*, edited by J. G. Ponterotto, J. M. Casas, L. A. Suzuki, and C. M. Alexander, 181–98. Thousand Oaks, CA: Sage.

Hunt, R. G., and B. Bowser. 1996. *Impacts of Racism on White Americans.* Sage Publications.

Karp, J. B. 1981. "The Emotional Impact and a Model for Changing Racist Attitudes." In *Impacts of Racism on White Americans*, edited by B. J. Bowser and R. G. Hunt, 87–96. Beverly Hills, CA: Sage.

Kivel, P. 1996. *Uprooting Racism: How White People Can Work for Racial Justice.* Philadelphia: New Society Publishers.

Kivel, P. 2017. *Uprooting Racism: How White People Can Work for Racial Justice.* Philadelphia: New Society Publishers.

Kleiman, S., L. B. Spanierman, and N. G. Smith. 2015. "Translating Oppression: The Effect of Sexual Minority Status on White Men's Racial Attitudes." *Psychology of Men and Masculinity* 16 (4): 404–15.

Kordesh, K., L. B. Spanierman, and H. A. Neville. 2013. "White Students' Racial Affect: Gaining a Deeper Understanding of the Antiracist Type." *Journal of Diversity in Higher Education* 6: 33–50.

Lee, B. X. 2019. *The Dangerous Case of Donald Trump: 37 Psychiatrists and Mental Health Experts Assess a President-Updated and Expanded with New Essays.* New York: Thomas Dunne Books.

Leonhardt, D., and I. P. Philbrick. 2018. "Donald Trump's Racism: The Definitive List, Updated." *New York Times*, January. https://www.nytimes.com/interactive/2018/01/15/opinion/leonhardt-trump-racist.html.

Lorde, A. 2012. *Sister Outsider: Essays and Speeches.* Berkeley, CA: Crossing Press.

Lydecker, J. A., R. R. Hubbard, C. B. Tully, S. O. Utsey, and S. E. Mazzeo. 2014. "White Public Regard: Associations Among Eating Disorder Symptomatology, Guilt, and White Guilt in Young Adult Women." *Eating Behaviors* 15 (1): 76–82.

McConnell, E. A., and N. R. Todd. 2015. "Differences in White Privilege Attitudes and Religious Beliefs Across Racial Affect Types." *The Counseling Psychologist* 43 (8): 1135–61.

Mekawi, Y., K. Bresin, and C. D. Hunter. 2016. "White Fear, Dehumanization, and Low Empathy: Lethal Combinations for Shooting Biases." *Cultural Diversity and Ethnic Minority Psychology* 22 (3): 322–32.

Merrilees, C. E., J. Katz, N. DuBois, and C. Grant. 2018. "White Female Bystanders' Responses to a Black Woman at Risk for Sexual Assault." *Violence and Victims* 33 (4): 739–54.

Neville, H. A., G. H. Awad, J. E. Brooks, M. P. Flores, and J. Bluemel. 2013. "Color-Blind Racial Ideology: Theory, Training, and Measurement Implications in Psychology." *American Psychologist* 68 (6): 455–66.

Ott, B. L., and G. Dickinson. 2019. *The Twitter Presidency: Donald J. Trump and the Politics of White Rage*. New York, NY: Routledge.

Paone, T. R., K. M. Malott, and J. J. Barr. 2015. "Assessing the Impact of a Race-Based Course on Counseling Students: A Quantitative Study." *Journal of Multicultural Counseling and Development* 43 (3): 206–20.

Pinterits, E. J., V. P. Poteat, and L. B. Spanierman. 2009. "The White Privilege Attitudes Scale (WPAS): Development and Initial Validation." *Journal of Counseling Psychology* 56: 417–29.

Poteat, V. P., and L. B. Spanierman. 2008. "Further Validation of the Psychosocial Costs of Racism to Whites Scale Among a Sample of Employed Adults." *The Counseling Psychologist* 36: 871–94.

Siegel, M. P., and R. T. Carter. 2014. "Emotions and White Racial Identity Status Attitudes." *Journal of Multicultural Counseling and Development* 42 (4): 218–31.

Sifford, A., K. Ng, and C. Wang. 2009. "Further Validation of the Psychosocial Costs of Racism to Whites Scale on a Sample of University Students in the Southeastern United States." *Journal of Counseling Psychology* 56: 585–89.

Smith, W. A., J. B. Mustaffa, C. M. Jones, T. J. Curry, and W. R. Allen. 2016. "'You Make Me Wanna Holler and Throw Up Both My Hands!': Campus Culture, Black Misandric Microaggressions, and Racial Battle Fatigue." *International Journal of Qualitative Studies in Education* 29 (9): 1189–209.

Soble, J. R., L. B. Spanierman, and H. Liao. 2011. "Effects of a Brief Video Intervention on White University Students' Racial Attitudes." *Journal of Counseling Psychology* 58: 151–57.

Spanierman, L. B., J. C. Beard, and N. R. Todd. 2012. "White Men's Fears, White Women's Tears: Examining Gender Differences in Racial Affect Types." *Sex Roles* 67: 174–86.

Spanierman, L. B., and N. L. Cabrera. 2015. "The Emotions of White Racism and Antiracism." In *Unveiling Whiteness in the 21st Century: Global Manifestations, Transdisciplinary Interventions*, edited by V. Watson, D. Howard-Wagner, and L. Spanierman, 9–28. Lanham, MD: Lexington Books.

Spanierman, L. B., and M. J. Heppner. 2004. "Psychosocial Costs of Racism to Whites Scale (PCRW): Construction and Initial Validation." *Journal of Counseling Psychology* 51: 249–62.

Spanierman, L. B., V. P. Poteat, A. M. Beer, and P. I. Armstrong. 2006. "Psychosocial Costs of Racism to Whites: Exploring Patterns Through Cluster Analysis." *Journal of Counseling Psychology* 53: 434–41.

Spanierman, L. B., V. P. Poteat, Y.-F. Wang, and E. Oh. 2008. "Psychosocial Costs of Racism to White Counselors: Predicting Various Dimensions of Multicultural Counseling Competence." *Journal of Counseling Psychology* 55: 75–88.

Spanierman, L. B., N. R. Todd, and C. J. Anderson. 2009. "Psychosocial Costs of Racism to Whites: Understanding Patterns Among University Students." *Journal of Counseling Psychology* 56: 239–52.

Terry, R. W. 1981. "The Negative Impact on White Values." In *Impacts of Racism on White Americans*, edited by B. J. Bowser and R. G. Hunt, 119–51. Beverly Hills, CA: Sage.

Todd, N. R., L. B. Spanierman, and V. P. Poteat. 2011. "Longitudinal Examination of the Psychosocial Costs of Racism to Whites Across the College Experience." *Journal of Counseling Psychology* 58: 508–21.

Turner, M. M. 2011. "Personality Factors, Multicultural Exposure, and Cultural Self-Efficacy in a Model of White Fear." Doctoral dissertation, University of Notre Dame. ProQuest Dissertations Publishing.

9

Media: Fox News, Racism, and White America in the Age of Trump

Kalemba Kizito

Abstract Kalemba Kizito looked critically at Fox and other news organizations. What Kizito reveals is that Fox News and its audience are distinct from the others. Its most popular programs are not news but commentaries by Sean Hannity and Tucker Carlson, which, by the way, do not make any claim to be accurate and fact-checked. A large segment of the Fox audience relies only on Fox for their news and commentary. All the other news networks, including MSNBC, reply extensively on multiple sources and fact-check their news and commentary. High proportions of non-Fox news audiences also use multiple sources for their news and commentary. Fox News operates in an information silo where its audience misses most domestic news and virtually all international news. Additionally, there is little distinction between news and commentary. The outcomes are predictable. For one, a significant proportion of White America, who constitute the Fox audience, has been led into a dangerous worldview where they believe that their freedoms and government have been taken from them.

Keywords Fox News · Media · Social media · Trump · Audiences · Fact-check

---

K. Kizito (✉)
Communication, Boston College, Boston, MA, USA
e-mail: Kalemba.Kizito@bc.edu

D. W. Austin and B. P. Bowser (eds.), *Impacts of Racism on White Americans In the Age of Trump*, https://doi.org/10.1007/978-3-030-75232-3_9

In 1987, Denzel Washington starred in the Richard Attenborough epic drama "Cry Freedom" set in late 1970s Apartheid South Africa. Although the term apartheid has been used in common reference to South Africa, it is universally applicable to skewed race relations and conditions of racial segregation that characterize all settler colonies. Like many of the movies and literary works set in immediate post-independent Africa, the plot lines revolve around asymmetrical race relations and the agitation for freedom by those under the colonial boot. Washington acts as the revolutionary Steve Biko, who led a Black Consciousness movement to inspire the zeal and courage among his people relegated to squalor conditions in townships. His primary nemesis is a White liberal newspaper editor Donald Woods (Kevin Kline), who uses his media platform to disparage Biko's efforts. About halfway into the movie, Washington's character Biko is killed by the White South African police, and the rest of the movie becomes a plot by Donald Woods to escape to Europe and tell Steve Biko's story. In his own words, former movie critic Roger Ebert says, "You know there is something wrong with the premise of this movie when you see that the actress who plays the editor's wife is billed above the actor who plays the Black leader" (1987). Biko and Woods' relationship is instructive of the relationship between the Black population in the United States and the predominantly White media through which Black life narratives are framed and disseminated. "Cry Freedom" is a film about "bold" liberal journalism partial to the Black plight. It is also a story about the mediated representation of Blacks and their stories through the lens of White media.

There are vivid parallels between the movie "Cry Freedom" and the state of events currently unfolding in the United States. The parallels are set around real-life social-political undercurrents whose master narrative under a South African apartheid system has constructed Black natives everywhere as second-class citizens. The death of Steve Biko has wide-ranging symbolic parallels with George Floyd's death and many other Black men and women both in South Africa and the United States who have died at the hands of an apartheid police system.[1] A particular scene in the movie that represents the real 1976 open massacre of school children in South Africa peacefully protesting White apartheid eerily captures some of the ugly scenes in the 2020 U.S. protests that have been broadcast around the world. These are scenes of Police and paramilitary personnel charging into protesters nationwide who are on the streets on account of the murder of George Floyd. But more poignantly, the collective fatal plight of Washington's character Steve Biko, Sandra Bland, George Floyd, and many more, is mediated by journalism

[1] Apartheid refers to an official policy of discrimination on grounds of race. While commonly associated with South Africa, I argue here, that it is universally applicable to Western racial segregation.

structured through the lens of a predominantly White press that dictates how society processes Black narratives.

The now famous phrase "I can't breathe" is a human dignity cry—a cry for freedom. Moreover, the film title's poignancy is in the fact that the Black native had no recourse to telling their own story.[2] What we know about Biko and Black South Africa had to be communicated through the "benevolence" of a White news editor who, not free from his own racial biases, also partially agrees with the apartheid state that Biko's ideology was dangerous ("your ideas are dangerous"). If even liberals are biased, it is reasonable to question if institutions run by Whites are incapable of ridding themselves of racial bias. "Cry Freedom" underscores the obstacles of challenging racist/apartheid systems in a media landscape that is principally White. This chapter offers critical commentary on the operational modality of mainstream broadcast news media predominantly owned, operated, and run by Whites and why it negatively impedes the cure of U.S. racism, and ultimately misinforms Whites and reinforces their racism.

Media Operational Model

Standing out among a vast scholarship on media behavioral analysis is the 1988 groundbreaking study *Manufacturing Consent: The Political Economy of the Mass Media* by Edward Herman and Noam Chomsky. The overarching contribution of this study is its propaganda model and the five original filters that anchor it. According to the model, filters in news media serve as gatekeeping mechanisms for what information or narratives can circulate in the public sphere and under what operating circumstances. The filters function as a bulwark against any serious challenge to the American status quo. Hence, it renders the media a part of the Whiteness apparatus that sustains the legacy of racism. There are five filters: First is the size, concentrated ownership, owner wealth, and profit orientation of the dominant mass-media firms. Second is advertising as mass media's primary source of income. The third is a reliance of the media on information provided by the government, business, and "experts" funded and approved by these primary sources and agents of power. Fourth, "flak" is a means of disciplining the media. Finally, "anticommunism" is used as a national religion and control mechanism. These elements interact with and reinforce one another.

[2] Cry Freedom echoes Gayatri Spivak's postcolonial critique in "Can the Subaltern speak?" Who tells the story of the marginalized is often an apparatus of the ruling/power elite?

The raw material of news must pass through these successive filters, leaving only the cleansed residue fit to print or broadcast. The premises of discourse and interpretation and the definition of what is newsworthy are fixed in the first place. They explain the basis and operations of what amount to propaganda campaigns. The elite domination of the media and marginalization of dissidents results from the operation of these filters. The process occurs so naturally that media news people, frequently operating with complete integrity and goodwill, can convince themselves that they choose and interpret the news "objectively" and based on professional news values (Herman and Chomsky 2005, 2).[3]

News and Racism in the Age of Trump

Racism is violence by commission or omission. It could take the form of overt acts or subtle acts, all of which result in physical or psychological harm to groups who are its targets. There is an extended glossary of terms in which the violence of racism is couched. They include White supremacy, White nationalism, Fascism, Islamophobia, domestic terrorism, extremism, and the list goes on. U.S. jurisprudence has recognized many of these terms in reference to hate crimes only since 1968 (U.S. DOJ 2019). According to the Southern Poverty Law Center, we have seen a spike in overt acts of racism with the Trump presidency.[4] Among their findings, the report notes:

> As the country continues to experience White nationalist terror, extremist ideas long believed outside of the realm of legitimate politics are penetrating deeply into the mainstream, spawning public policies that target immigrants, LGBTQ people, and Muslims. The Trump administration installed members of hate groups into government—particularly those with anti-immigrant, anti-Muslim, or anti-LGBTQ animus—and put in place highly punitive policies that seemed unthinkable just a few short years ago. These political moves will

---

[3] In a recent 2018 interview, Herman was asked how the propaganda model could be improved. He intimated that: We modified it in 2002 to include free-market ideology as an important ideological premise, complementing anticommunism. We could improve it by spelling out in much more detail how elite divisions and local factors affect the media and allow dissent more space … We could update on the growing command of militarism and war and the consequent growth in war propaganda (E.S. Herman 2000, 98). The recent uptick in racism during the presidential term of Donald Trump is also an example of how Whites' harness the media to propagate racism. The model can improve by accounting for race as a critical filter, given its prevalence in elite media.

[4] Founded in 1971, the Southern Poverty Law Center is a nonprofit legal organization that works for the promotion of Civil rights and social justice. They monitor and track domestic hate groups in the U.S., and actively pursue litigation in the area of Civil Rights. Their work is cited in government reports as well as in scholarly work.

> far outlast this administration, as Trump and his allies in the U.S. Senate have pushed through hundreds of new federal judges, many of whom are hostile to civil rights concerns and will serve for decades. (SPLC Report 2020, 7)

As the report notes, under the Trump administration, there has been an explosion of open racist attacks against minorities compounded by government officials' appointments sympathetic to White supremacy causes. Although racism is part of the American social fabric, it should be underscored that with Trump in office, the level of avowed racism is shocking as we journey into this new century. The report points to a Guardian newspaper article that connects Trump's Facebook use of the term "invasion" in more than 2000 ads. This language is a partial inspiration for increasing racist acts of violence, such as the August 2019 El Paso shooting (Aratani 2019). Trump has emerged as an avowed spokesperson for White racists since his presidential campaign (Phillips 2017). One aspect the Southern Poverty Law Center report does not address is media complicity—especially mainstream cable news media—in Trump's rise and the spike in racist violence. To illustrate the role of media culpability in the current political climate, let us consider Fox News on one end of the spectrum. MSNBC and CNN are on the other end. The three primary players in cable news have tag-teamed to inflame racial polarity during the Trump Presidency.

Fox News and the Racial Filter

The racial filter universally implicates media organizations based on ownership, management, and demographic composition; however, there is a stark difference between Fox News and the rest of the media landscape. Among the biggest headlines of 2020 garnering domestic and international news attention have been the current Covid-19 global pandemic and the transnational protests sparked by the lynching of George Floyd in Minnesota. Taking cues from Trump, Fox News, owned by Rupert Murdoch of the News Corp empire, covered both these events with rhetoric that depicted open racism condoned by the news organization and its predominantly White audience. Trump appeared several times on TV using the phrase "the Wuhan/Chinese virus" even against diplomatic protestations not to do so (Mangan 2020). Fox News ran with this racist framing. A Media Matters article details the number of times both bigoted phrases were used by Fox News anchors—144 times—including 43 times by leading prime time anchor Tucker Carlson (Savillo 2020). On the general topic of demographic change animating the radical right, the Southern Poverty Law Center report also named Tucker Carlson and his colleagues as propagators of radical right racism (Mangan, 10–11). Concerning Black Lives Matter protests after George Floyd's police murder, Fox News used disparaging language about protesters, calling them looters and left-wing rage mobs (Carlson 2020). This was not just for the 2020 protests. A study by Colleen Mills (2017) looked at the Black Lives Matter protests that erupted after Michael Brown's killing, a Black man whose body was left on the street for close to four hours. She found that Fox News uses disparaging language against protesters to deflect the criminal justice system's racism. Mills characterizes Fox News coverage as sanctioned "through a prism of underlying ideologies, specifically those of White supremacy and racial capitalism" (40). Ideology is used as a filter to frames what organizations report as news.

Fox News channel reporter William La Jeunesse wrote about the civil rights protests under the headline—"George Floyd protests could be most expensive civil *disturbance* [emphasis mine] in US history, experts say." In the same breath, he referred to the 1992 Los Angeles protests as the costliest civil *disorder* [emphasis mine] event to date in United States (La Jeunesse 2020). It did not matter that people worldwide rallied in civic solidarity with U.S. protesters to demand an end to racist policing. La Jeunesse framed the protests dismissively as "disturbances" and "disorderliness." Audiences partial to Fox News's ideological bend agree with this frame instead of questioning police methods. This kind of journalism deliberately prevents any critiques of racist policing tactics. In turn, this approach contributes to the intransigence of a segment of the White population that continues to dismiss the veritable impacts of racist policing on Blacks and other minorities. Fox News has carved out a niche reporting that capitalizes on race-baiting journalism for its predominantly White conservative Republican audience. However, CNN and MSNBC are also complicit in shoring up the other side of the wall that subtly greases the nation's basest instincts.

MSNBC

On the cover of his recent book *Hate Inc.: Why Today's Media Makes Us Despise One Another*, veteran journalist Matt Taibbi (2019) aptly puts the pictures of Fox News firebrand Sean Hannity and MSNBC's star anchor Rachel Maddow standing back to back. They are each facing in the opposite direction. His choice of the cover illustration is revealing of some simple truths. American cable news is dominated by a cast of White characters, and that as audiences, we align ourselves with one of two sides (at least as far as cable news goes). He writes:

> It boggles my mind that people think they're practicing real political advocacy by watching major corporate TV, be it Fox or MSNBC, or CNN. Does anyone seriously believe that powerful people would allow truly dangerous ideas to be broadcast on TV? The news today is a reality show where you're part of the cast: America vs. America, on every channel. (Taibbi 2019, 42)

In other words, the journalistic spectacle that is cable news does not function to produce meaningful change. Taibbi acknowledges that while Fox News is a mouthpiece for White America's basest instincts, "people who watch MSNBC, meanwhile, are tuning in to receive mega-doses of the world's thinnest compliment, i.e., that they're morally superior to Donald Trump and his followers. The network lately has become a one-note morality play with endless segments about Michael Flynn, Michael Cohen, and Paul Manafort" (Taibbi 2019, 140). The past four years on MSNBC have been an open-ended fixation with Trump's open racism. By focusing on Trump, the liberal left excuses itself from reflecting on their own complicity in sustaining America's worst attitudes.

Decolonial scholars Tuck and Yang have characterized MSNBC's approach as "moves to innocence" (Tuck and Yang 2012). Media scholar Katherine Bell (2019) similarly argues that when MSNBC coverage over fixates on overt individual racism, they play it safe from confronting the uglier institutional side of American racism that has been and still continues to thrive. Commenting on the respective mandates of both networks, Taibi notes that: "The coverage formula on both channels is to scare the crap out of audiences, then offer them micro-doses of safety and solidarity, which come when they see people onscreen sharing their fears. There is a promise of reassurance that comes with both coverage formulas" (Taibbi 2019, 141).

An example of "moves to innocence" coverage was apparent in the CNN and MSNBC reporting on the George Floyd protests. In their reporting, they did not use disparaging language; they provided ample coverage of the protests. However, what was lacking was a proper context of America's history that continues to produce the sad spectacle of George Floyd type state-sanctioned murders. Let us consider the following MSNBC program. A weekend anchor, Ali Velshi run a segment on July 26, 2020, whose headline was "Swing State Voters Disagree on Whether Trump is a Racist." In this segment, he parades voters from Pennsylvania and asks if they believed Trump was a racist. As stated by the propaganda model, in the first instance, the acceptable frame of debate on racism is whether Trump is a racist. For a pedestrian viewer, a debate is seemingly taking place. But the parameters of the debate have been already filtered to narrow its terms.

Reducing racism in America to Trump totally misses the point that America has long been racist. By default, it frames individual racism as outside of the larger history of the nation's racism. This selective framing is a disservice to viewers. In a second instance, on July 30th, Velshi ran a segment titled, "The Withdrawal of troops From Germany is Being Called a Gift for Putin. Here's Why." In this segment, MSNBC softly pushed a war stance using the Putin boogeyman. But that is beside the point. The real discussion ought to be about how American war adventures abroad encourage domestic terror in the hands of law enforcement back home. Such a headline may not be critiqued easily by MSNBC's loyal audience trained to react negatively at the mention of Putin.

A recently released HBO documentary, "After truth: Disinformation and the cost of fake news," makes interesting observations. There are negative impacts of living in a weaponized information environment (Stelter 2020). Predominantly focused on Fox News and its audience, the film shows how news organizations can exploit audience ignorance, anti-intellectualism, and pre-existing fear among Whites to propagate extreme ideological programming and conspiracy theories. The fake news phenomenon cheer-led by the president exploited a cultural environment in which audiences are already splintered along ideological lines. Such a news environment does not favor objectivity, but rather a news model that satisfies fidelity to pre-constituted world views. In the end, the film makes a crucial underlying point about White America, fake news, and propaganda; It is less about which side is more extreme. Whether it is fake news or propaganda, existing cable news models suggest that news organizations present information that their targeted audiences are already pre-disposed to believe. Recently, Comedian Chris Rock in a *New York Times* article summed it well when he said that "Put it this way: Republicans tell outright lies. Democrats leave out key pieces of the truth that would lead to a more nuanced argument. In a sense, it's all fake news" (Itzkoff 2020). However, circling back to the beginning of this essay, Donald Woods promises that individual White journalists will rise to the occasion. Just as Woods realizes his complicity in the apartheid system and switches sides, that possibility remains open for White journalists to quit mainstream media organizations and pursue individual journalism that is more ethical.

Implications

While race is maintained as a structuring lens through which the Other is seen in the media, Mr. Woods, illustrates the possibilities for White-owned news organizations to overcome their biases in favor of objective reporting. After meeting and talking with Biko, walking with him through the Townships, and seeing the conditions of the natives, Woods adopts a different journalistic stance. Similarly, the objective reporting on Covid-19 and the coverage of protests against racist policing by CNN and MSNBC, and other news organizations, call to mind Mr. Woods. White journalists can and must objectively show a broader frame informing and exposing systemic disparities and the violence attending to people of color.

In sharp contrast, Fox News, cheer-led by President Trump, stood unrivaled in its promotion of racially charged reporting that whipped-up anti-Black bias against the protesters. Fox News race-bait journalism suggests that a segment of White America will continue to be badly misinformed. For instance, when other stations were doing programming about the science of Covid-19, Fox News hosted programs that called it a hoax created by the Democratic Party. In one segment, Eric Trump called social-distancing a Democratic Party electoral strategy. An issue like wearing masks that the CDC says is our best weapon in combating Covid-19 spread got plenty of cynical and conspiratorial coverage. Several anchors voiced their suspicion about wearing masks.[5] A lawsuit filed by Washington League for Increased Transparency and Ethics (WASHLITE) accuses Fox News under the Washington State Consumer Protection Act (CPA). They sought both penalty and injunctive relief for misleading the public and downplaying the seriousness of Covid-19. Writing for Forbes, Sullivan (2020) notes how 74 journalism professors in an open letter to Fox News Channel chastised it for endangering the public with its politicized coverage disparaging the virus's seriousness. Evidence of how this was threatening to the public and Republicans themselves was in full display at Trump rallies with no social distancing or mask-wearing. After attending one of the Trump rallies, former Republican presidential candidate Herman Cain contracted and died of Covid-19 (Law 2020; Sullivan 2020).

Moreover, the danger to Fox News' Republican audience is complicated because they are more likely, according to a Pew report (Gramlich 2020), to seek their information just from Fox News. Democrats and Independents draw information from multiple sources. A recent Pew Research article on

[5] How Fox News helped turn masks into another culture war flashpoint. https://www.mediamatters.org/coronavirus-covid-19/how-fox-news-helped-turn-masks-another-culture-war-flashpoint.

audience commitment found that those committed to Fox News believe they have the most *trusted* news source. In contrast, non-Fox News audiences are not limited to MSNBC or CNN. Therefore, the typical conservative Republican is more likely to be misinformed due to a constricted media diet (Jurkowitz et al. 2020).

The idea that Republican voters are ill-informed remains consistent in research findings and reports on audience information. A Quora contributor in a 2016 Forbes article that examined the "Fox News Effect," cites William Poundstone who presents it this way:

> There's a lot that goes on in the world that doesn't easily fit the Fox template. There are important stories that don't make anyone angry, prove liberals are evil, or otherwise carry an emotional punch. Fox viewers get less of them. Fox News is like an all-you-can-eat buffet, serving up red meat. A more balanced diet might be healthier in the long run. (Poundstone 2016)

Along the lines of misleading journalism, In an article by the Poynter institute—an organization dedicated to ethical integrity in journalism—the author writes, "People who watch MSNBC and CNN exclusively can answer more questions about domestic events than people who watch no news at all. People who only watch Fox did much worse. NPR listeners answered more questions correctly than people in any other category" (Beaujon 2012). This point was also reiterated in Bartlett (2015) in a detailed study that Fox viewers are less informed. However, what is more indicting in Bartlett's article, is that it goes beyond simply telling us what most media analysts have said about Fox's audience. It tells us that Fox lies to its audience. This is why Fox viewers are more likely to believe that Obama was not born in America. It also details how Republican high-ranking members also complain about Fox, which constitutes the party's ideological wing. A negative downside for Republicans and the rest of the nation is that the news organization forces its politicians to wager on the extreme and lunatic! Therefore, whether or not Donald Trump is a racist, or whether Fox viewers are racist or not, the fact is that by dancing solely to the tune of Fox, they are forced to mime in mimicry of the racist ideological undertones that percolate the Fox airwaves.

Overall, as long as viewers remain glued to Fox News and are unwilling to expand beyond that echo chamber, they will continue to be less-informed, misinformed, and pose a danger to the rest of society. As shown in the preceding paragraphs, Fox News and President Trump led the racist charge of framing Covid-19 as a "Chinese virus." The spikes in anti-Asian sentiments violence were attributed to racist views on Fox News and President Trump's racist rhetoric. The unprecedented spike in violence summarized the costs of

Fox News reporting in the age of Trump in rhetorical and symbolic violence among Whites afflicted with racism. Afflicted with this disease, perhaps the biggest impact on Whites is that their capacity for effective civic engagement is curtailed severely.

Conclusion

On January 13, 2021, Donald Trump became the first U.S president to be impeached twice, and with ten House Republicans joining the 232–197 vote, making it a bipartisan impeachment. His impeachment followed his direct role in the incitement of a January 6 insurrection that saw White supremacists storm the Capitol with Congress in session in a failed coup attempt that saw five people die. The DC spectacle—of civic decadence—speaks volumes about the ongoing role that White supremacy plays in American Society and is a compelling end to the Trump Presidency. The Trump era has for lack of a better phrase, provided a "shocking" epiphany for many Whites about that which Black people, Indigenous and other minorities have always known—that White supremacy is a danger not just to them, but to the very fabric of the American social experiment. *The Washington Post* reported on details of an intelligence briefing that reflects what Fox News and President Trump's racist rhetoric helped build in the last four years.

"Be ready to fight. Congress needs to hear glass breaking, doors being kicked in, and blood from their BLM and Antifa slave soldiers being spilled. Get violent. Stop calling this a march, or rally, or a protest. Go there ready for war. We get our President, or we die. NOTHING else will achieve this goal" (Barett and Zapotosky 2021).

This excerpt redacted from an FBI report on Jan 5th, a day before the raid on the joint congressional session sums up what racism stretched to its logical conclusion yields—At its worst, racism draws blood. Any attempts by Fox News or by President Trump to save face after accommodating and encouraging the ecology of ideas responsible for spikes in White supremacist violence in the past four years, will hopefully inspire a deep introspection for White America to swallow a bitter pill—that racism deeply haunts this settler society. Whites must search deep within themselves, and can no longer hide behind first amendment rights, patriotism, or the American flag to accommodate and continue to sustain racism. The fall-out of the last four years in race relations will continue to be felt long after President Trump is out of office. Currently, several corporate donor organizations are lining up to cut funding for any Republican politicians complicit in encouraging the coup.

Many have cut ties with the Trump organization. Given the business model of American cable news organizations, we will see if enough pressure will be brought by donors to hold Fox News accountable as a news organization.

Discussion Questions

Q1: What are the five media filters and how do they serve as gatekeeping mechanisms for what information or narratives can circulate in the public sphere and under what operating circumstances?

Q2: What makes the Fox News audience distinct from other new audiences such as MSNBC? How does this impact the mostly White Fox News audience compared to other audiences?

Q3: What was the relationship between Fox News and Donald Trump during most of his presidency? How did that impact did that relationship have on the most loyal Fox viewers?

References

Aratani, L. 2019. "Invasion and Fake News: El Paso Manifesto Echoes Trump Language." *The Guardian*, August 5. https://www.theguardian.com/us-news/2019/aug/05/el-paso-shooting-suspect-trump-language-manifesto.

Barett, D., and M. Zapotosky. 2021. " FBI Report Warned of 'War' at Capitol, Contradicting Claims There Was No Indication of Looming Violence." *The Washington Post*, January 14. https://www.washingtonpost.com/national-security/capitol-riot-fbi-intelligence/2021/01/12/30d12748-546b-11eb-a817-e5e7f8a406d6_story.html.

Bartlett, B. 2015. "How Fox News Changed American Media and Political Dynamics [Scholarly Project]." *SSRN*, June 3. https://papers.ssrn.com/sol3/papers.cfm?abstract_id=2604679.

Beaujon, A. 2012. "Survey: NPR's Listeners Best-Informed, Fox Viewers Worst-Informed." Accessed 18 December. https://www.poynter.org/reporting-editing/2012/survey-nprs-listeners-best-informed-fox-news-viewers-worst-informed/.

Bell, K.M. 2019. "This Is Not Who We Are:" Progressive Media and Post-Race in the New Era of Overt Racism. *Communication Culture & Critique* 12 (1): 1–17.

Carlson, Tuck. 2020. "Tucker Carlson Tonight: The Rise of Left-Wing Mobs." Last Modified June 9. https://www.youtube.com/watch?v=fv6Y3a4BV6E.

Ebert, Roger. 1987. "Reviews: Cry Freedom." Last Modified November 6, 1987. Accessed 2020. https://www.rogerebert.com/reviews/cry-freedom-1987.

Gramlich, J. 2020. "5 Facts About Fox News From." Accessed 8 April. https://www.pewresearch.org/fact-tank/2020/04/08/five-facts-about-fox-news/.

Herman, E.S. 2000. The Propaganda Model: A Retrospective. *Journalism Studies* 1 (1): 101–112.

Herman, E.S., and N. Chomsky. 2005. *Manufacturing Consent*. New York: Pantheon.

Itzkoff, D. 2020. "Chris Rock Tried to Warn Us." *New York Times*, September 16. https://www.nytimes.com/2020/09/16/arts/television/chris-rock-fargo.html.

Jurkowitz, M., A. Mitchell, E. Shearer, and M. Walker. 2020. *US Media Polarization and the 2020 Election: A Nation Divided*. Pew Research Center.

La Jeunesse, W. 2020. "George Floyd Protests Could Be Most Expensive Civil Disturbance in U.S History Experts Say." Accessed 18 September. https://www.foxnews.com/politics/george-floyd-protests-expensive-civil-disturbance-us-history.

Law, Tara. 2020. "Eric Trump Claims Social Distancing Is a Democrat 'Strategy' and COVID-19 Will 'Magically' Disappear After Election." *Time*, May 18.

Mangan, D. 2020. "Trump Defends Calling Coronavirus 'Chinese Virus'—'It's Not Racist at All." Retrieved from. Accessed 15 March. https://www.cnbc.com/2020/03/18/coronavirus-criticism-trump-defends-saying-chinese-virus.html.

Mills, C.E. 2017. Framing Ferguson: Fox News and the Construction of US Racism. *Race & Class* 58 (4): 39–56.

Phillips, Amber. 2017. "'They're Rapists:' President Trump's Campaign Launch Speech Two Years Later, Annotated." *The Washington Post*, June 16.

Poundstone, W. 2016. "A Rigorous Scientific Look Into the 'Fox News Effect'," July 21.

Savillo, R. 2020. "Fox News Pushes Racist and Xenophobic Language for Coronavirus," June 15. https://www.mediamatters.org/coronavirus-covid-19/fox-news-pushes-racist-and-xenophobic-language-coronavirus.

SPLC Report. 2020. "The Year in Hate and Extremism 2019," 2020. https://www.splcenter.org/news/2020/03/18/year-hate-and-extremism-2019.

Stelter, Brian. 2020. "After Truth: Disinformation and the Cost of Fake News." https://www.hbo.com/documentaries/after-truth-disinformation-and-the-cost-of-fake-news.

Sullivan, Bryan. 2020. "Fox News Faces Lawsuit for Calling COVID-19 A 'Hoax'." *Forbes*, April 10.

Taibbi, Matt. 2019. *Hate Inc.: Why Today's Media Makes Us Despise One Another*. New York: OR Books.

Tuck, E., and K.W. Yang. 2012. "Decolonization Is Not a Metaphor." *Decolonization: Indigeneity, Education & Society* 1 (1).

U.S. DOJ. 2019. "Hate Crime Laws." Last Modified March 7, 2019. https://www.justice.gov/crt/hate-crime-laws.

10

Sports: Racism and Sports—Fear of a Black Planet

Scott N. Brooks, Stacey M. Flores, and Jorge Ballesteros

Abstract College and professional football and basketball are areas in American life where there is supposedly a Black–White win–win. Black athletes gain fame and fortune, and White fans enjoy extraordinary athletic contests. Scott Brooks and colleagues provide a case study of the racism undermining this win–win, which illustrates how cultural racism is deplored and has effects on Whites. They show not only what happens to Black athletes who get out of line, but also what happens to White staff and faculty who side with the Black athletes or who get in the way of their punishment. Suppression of Black protest is costly to Whites as well.

Keywords Sports · Black athletes · Racism · Boycott · Students · Sports fans · Slavery

Sport in the United States is frequently said to be an exception, a space of racial progress, but this perception often references Black athletes' success, not looking at all sports nor the sports industry beyond the athlete. However, a critical look at sports reveals that not only are they a mirror of society's racism, but perhaps the purest remaining remnant of slavery (Rhoden 2006). Sports remain very segregated (racially and economically); there is a strict color line

S. N. Brooks (✉) · S. M. Flores · J. Ballesteros
Arizona State University, Tempe, AZ, USA
e-mail: scott.n.brooks@asu.edu

D. W. Austin and B. P. Bowser (eds.), *Impacts of Racism on White Americans In the Age of Trump*, https://doi.org/10.1007/978-3-030-75232-3_10

that divides labor/athletes from the C-suite and ownership. The color line is reflected in what we see and in the words of professional sports executive managers and owners who have expressed White supremacist attitudes—that Whites are better than Blacks.

These managers have been explicit in their racism from captured private comments to unapologetic interviews on network television. In 1987, CBS's "Nightline" featured a story on racial progress in baseball to celebrate the 40th anniversary of Jackie Robinson's major league debut. Al Campanis was asked to be on the show because of his relationship with Jackie Robinson; he and Robinson were teammates and sometimes roomed together on road trips. Campanis, then vice president and general manager of the Los Angeles Dodgers, was asked a series of questions by Ted Koppel. "Why is it that there are no Black managers, no Black general managers, no Black owners... Is there still that much prejudice in baseball today?" (Weinbaum 2012). Campanis shocked many with his response: "No, I don't believe it's prejudice. I truly believe that they may not have some of the necessities to be, let's say, a field manager, or perhaps a general manager." Koppel was visibly surprised and responded, "sounds a lot like the garbage we heard 40 years ago." The immediate conclusion reverberated; racism remained deeply entrenched in baseball. Two days later, Campanis lost his job, even though he had helped the Dodgers reach the World Series four times. He was defended by former teammates and players, Black and White, who said that Campanis was not a racist. Campanis sits as a cautionary tale for White sports executives and managers: be careful of what you say when you talk about race.

Marge Schott, former owner of the Cincinnati Reds Major League Baseball club from 1984–1999, referred to some of her team's Black players as "million-dollar niggers" (Associated Press 1999). In 2014, Los Angeles Clippers' owner Donald Sterling revealed his sense of White superiority on a phone recording released by his mistress. In the message, Sterling told her she should not be seen associating with Black people. She challenged his position: "Do you know that you have a whole team that's Black, that plays for you?" Sterling responded, "Do I know? I support them and give them food and clothes and cars and houses. Who gives it to them? Does someone else give it to them? Who makes the game? Do I make the game, or do they make the game?" (2019). Sterling had been the Clippers' owner for thirty-three years.

These conversations went public, and while Campanis was defended, Schott and Sterling doubled down when questioned. Schott did not deny having anti-Black attitudes or saying the slurs, and Sterling, unable to avoid distinguishing himself from "they" (Blacks) did not fare much better. These events are racial moments, and all went "viral" in the media. They reveal how the consequences of White supremacy benefit and penalize Whites differently. White elites have the most advantages and gain the most from exercising racism. Campanis, an executive at the time, lost his job and became a social pariah. Schott faced no real backlash and maintained ownership. Sterling had a long history of racist actions and poor relationships with his Black staff and players and was forced out of ownership but made a $1,987,500 profit or a 15,900% return (Babb 2014). In the end, the public outcries following these moments have not advanced the bar of racial progress. They are described as the acts of individuals who exhibited race prejudice and feelings of superiority; they say nothing about the racist institutions for which these individuals and others like them are responsible. Is anyone hurt by their attitudes? This chapter explores this question and the implication of the answer for the age of Trump.

Positions in White Supremacy

From nineteenth-century slave uprisings to taking a knee during the national anthem, White supremacist attitudes and White fear have co-existed in the United States in managing Black people. In his autobiography, Frederick Douglass wrote about White fear during slavery in the story of Demby, a slave killed by an overseer. Demby took a few lashes of the overseer's whip and then ran into a creek, refusing to get out. He thought that he had saved himself. The overseer, "Mr. Gore," gave the ultimatum: come out or be shot. When Demby still refused, the overseer shot and killed him. Mr. Gore explained his actions to the slave master, professing that Demby "had become unmanageable." He believed that if one slave refused to be corrected and escaped with his life, the other slaves would soon follow suit, and the ultimate result would be slave revolt and White enslavement and death. As an overseer, it was his job to protect the slave owner's interests and the social order.

This story illustrates how White supremacy as a racial order has been maintained through class-based practices that define the racial order and carry out the order in everyday practices. Whites are not a monolith; they have different positions in the racial order and do not all work in the service of White supremacy. Douglass recalled a good White woman, his slave master's wife, who taught him to read but then was told to stop teaching him by his slave master.

> The fatal poison of irresponsible power was already in her hands and soon commenced its infernal work. That cheerful eye, under the influence of slavery, soon became red with rage; that voice, made all of sweet accord, changed to one

> of harsh and horrid discord; and that angelic face gave place to that of a demon . . . Mr. Auld found out what was going on, and at once forbade Mrs. Auld to instruct me further, telling her, among other things, that it was unlawful, as well as unsafe, to teach a slave to read. To use his own words, further, he said, "If you give a nigger an inch, he will take a mile. A nigger should know nothing but to obey his master—to do as he is told to do. Learning would *spoil* the best nigger in the world. Now," said he, "if you teach that nigger (speaking of myself) how to read, there would be no keeping him. It would forever unfit him to be a slave. He would at once become unmanageable, and of no value to his master. As to himself, it could do him no good, but a great deal of harm. It would make him discontented and unhappy." (Douglass 1845, 28–9)

With phrases like "fatal poison of irresponsible power," Douglass described a social psychological impact of slavery and racism: it poisons Whites. He noted an attitude change in his slave master's wife, from sweet to harsh and horrid and from angelic to demonic. Whether she accepted her husband's racist ideas or feared opposing him, she became a follower. How does considering Blacks to be inferior—and even worthy of death—cost Whites? This is a question not just for non-elite Whites, it is one for Whites who manage and fill in between White elites and all others— like the overseer and the slave master's wife.

We generally consider some racists to have social, economic, and personal needs for White supremacy. Likewise, conformists go with prevailing racial attitudes. Then, we think of anti-racists as like historical abolitionists, people actively engaged in thwarting and fighting against White supremacy. In contrast, Douglass' excerpts above suggest four roles played by antebellum Whites: one by the slave master, another by the overseer, and two roles played by the master's wife. We need to pay closer attention to the overseer and master's wife since they played "middleman" roles.

Most Whites actively police and guard social and economic borders between people of color and the White elite and doing so have differing effects and create social costs for themselves. We break down Whites into groups to acknowledge their diversity in social position. For example, Bonacich described White ethnics as a "middleman minority" in terms of status and economic position (1973). Occupying an intermediate position, they are described as neither owners nor executives of a large corporation nor the bottom level workers; rather, middle managers and small businesspersons. Middle(wo)man minorities are significant because they fill the space between the elites and masses of workers while posing little threat to the overall power structure.

To illustrate the costs of racism to all involved—including Whites, we will use a "big event"—the University of Missouri's 2015 football protest—as a case study. Attention will be paid to different White social positions, connections, and responses to the protest and the impact of events in defining and protecting the racial order. Below we consider three moments from this event to illustrate racism in sport, the unequal positions of Black athletes and White sport franchise managers/executives and owners, and its impacts on Whites.

A Big Event

The 2015 University of Missouri (MU) football protest was a big event with three general and unequal impacts on Whites. (1) Whites lost jobs; (2) Whites were punished; and (3) the historically White university has not recovered financially or socially, nor does it look like it will recover soon. The protest made Whiteness visible, and several actions within the big event illustrated cultural, institutional, and individual racism (Bowser 2017). Outside of Missouri, the campus protest served as a warning to Whites who may not have seen the threat of Black equality and liberation—a takeover—that must be managed and thwarted.

Intercollegiate sports are the university's front porch (Zimbalist 1999), and big wins are signs of a school's excellence. Football and basketball are the only revenue sports for colleges and universities; football pays for the rest of college sports. Admissions grow as more Whites become interested in being connected to the sport's reputation of a school. Sport is a space of interest convergence: sport uses Black labor to benefit historically White schools and colleges, multinational corporations, White professional franchise owners, and fans. A disproportionate number of athletes are Black, and the coaches, athletic directors, boosters, and school presidents are White.

The support of boosters and White students at MU was used to maintain systemic and institutional racism. White students have a cultural connection to the sport as a part of their historically White institution. Many White students and fans feel pride watching Black student-athletes, thinking little of the race-based caste system, a labor relationship controlled by White interests, wealth and power, and a sense of superiority (Rhoden 2006).

What Started the Protest

MU places considerable value on its football program and athletes. In 2012, the university joined the Southeastern Conference (SEC)—America's premier college football conference. While this bump in its football status required capital investments, including its stadium's renovation, the investments would pay off. New luxury boxes would accommodate more big-money supporters. Crowds of nearly 70,000 fans were raucous in their support of the football program, willing to show up for games in freezing rain or sticky heat. Players were treated like celebrities. They were on ESPN, recognized in public, and given free meals at local restaurants and gear through the athletics department. As first-year students, football players were housed primarily in the "athlete dorm," where they could experience college life's sexual freedom. Young women trolled the hallways looking for well-known recruits, as well as anyone on the team. Players walked freely with only towels around their waists, expecting to get lucky.

However, things were very different on campus. In classes, Black student-athletes were viewed with awe and skepticism by students and faculty who considered them superstars and prima donnas, privileged, and undeserving students (Comeaux 2012). In contrast, Black students (non-athletes) received only the stigma of Blackness: the racism that said they were less than Whites, and nothing could be done to change this. One of the authors, who was faculty at Missouri during this time, wrote about what he learned from speaking and interacting with students, including many football players, and closely observed the campus social climate (Brooks 2016). MU's new standing in intercollegiate football had created a new dynamic on campus.

On August 9, 2014, Michael Brown was murdered by police in Ferguson, a suburb of St. Louis, Missouri, less than 120 miles from the University of Missouri in Columbia. Led overwhelmingly by Black students, silent protests in response to Brown's murder began in the fall of 2014. Signs read, "Ferguson is here." Students said clearly in interviews and public forums, "Racism lives at the University of Missouri, and so do we." They educated those who would listen: Michael Brown's killing was not an isolated event—any Black person could suffer Michael Brown's fate, and police brutality and institutionalized racism were everywhere. There was more than one voice within the Black students' response, and they did not all agree. They formed as many as 7–10 Black student groups, with different foci and approaches to change. One of these organizations was Concerned Student 1-9-5-0 (CS1950), named for the year the University (MU) admitted its first Black student.

Initially, many Whites were surprised by the campus activity; they were used to student demonstrations, but not by Black students *en masse*. Early on, Chancellor Loftin held town hall meetings, invited students to have chats, and met with student organizations to discuss how to make MU better for Black students. But in little time, some White students expressed annoyance and complained about being inconvenienced (Jo 2015). Proprietary claims to the campus, grounded in Whiteness, made it easy to characterize Black student protesters as whiners and dramatists, trying to make a "mountain out of a mole hill." At the outset of the student demonstrations, Loftin publicly set out an ambitious agenda to end racism on campus but stopped way short of doing so.

Frustrated by Loftin's inaction, CS1950 took their demands to the systemwide President, Tim Wolfe. The group's leader, Jonathan Butler, told a journalist, "All we get is emails and empty promises. And we're here to say we're not going to be OK with just emails or empty promises anymore." Rather than addressing Black students' concerns, Wolfe refused to learn the concepts of racism, refused to meet with students, and consider working to end racism. He held the line, even with his job on the cutting block.

Football Joins In

CS1950 deepened its commitment and organization. A grad student leader, Jonathan Butler, took the courageous and self-sacrificial step of going on a hunger strike. On November 7, the sixth day of his hunger strike, Butler and CS1950 met with many of the Black members of the football team. They were moved and drawn to his life-threatening commitment to a just cause. Black MU football players decided to support the cause and called for the ousting of President Wolfe. Thirty Black football players took a photo with Jonathan Butler and tweeted:

> The athletes of color on the University of Missouri football team truly believe. "Injustice Anywhere is a Threat to Justice Everywhere." We will no longer participate in any football-related activities until President Tim Wolfe resigns or is removed due to his negligence toward marginalized students' experiences. WE ARE UNITED!!!!!

Until Wolfe resigned or was fired, they would boycott all football activities in solidarity with Butler. This included practice and their upcoming game against Brigham Young University.

The Backlash

Whites responded in four ways to the football players' protest and the idea that MU was not a safe, equitable, and supportive place for Blacks. Anti-racists Whites talked about Black students' rights and joined the Black protesters. Other Whites did not seem to care one way or another; they worried most about being inconvenienced. They were potentially like the overseer and master's wife—they were likely to cooperate with which/whoever serves their interests. Still, other Whites like the master wanted to push Blacks back and stamp down protests; they talked about Blacks' lack of gratitude for attending MU. These White students responded to Black student protests with counter-protests and no empathy. This last group had the most influence and the biggest platforms. They spoke against what the athletes did and were unmoved by Black students' feelings of being neglected and mistreated. Fortunately, they did not act physically against the Black students; they spoke from their perches. While we have named these groups, people moved between groups as it serves their interests, and situations changed.

The Overseers

MU's athletic director, Mack Rhoades, and Coach Pinkel held a press conference to announce that they would support their football players. Within 48 h, President Wolfe called a press conference and announced that he was stepping down as systemwide President. CS1950 claimed a victory, and Jonathan Butler ended his hunger strike. Missouri Gov. Jay Nixon thanked Wolfe for resigning, claiming it was "a necessary step toward healing and reconciliation." Someone had to be sacrificed, and it needed to be done promptly to restore order: the players said they were ready to play following Wolfe's resignation (Bogage and Reiss 2015). Mack Rhoades and Gary Pinkel were in the role of overseers but did not realize it. They were active in resisting anti-Black racism by siding with the football players against the University president. But they did not realize that the role of being in the middle is a tough line to hold.

Melissa Click, a White Professor, was charged with a misdemeanor after blocking a journalist from filming CS1950. In earlier protests, Professor Click got in the face of a police officer and cursed at him. She was let go from her lecturer appointment—the first White casualty. And along with Professor Click, Janna Basler, an administrator with Greek life, was placed on administrative leave, relieved of her duties, and investigated for her actions in protecting CS1950 (Yan and Stapleton 2015)—the second White casualty. Both were seen as overstepping their job duties: protecting Black students was political. They were supposed to be politically neutral or go along with the White normative order.

Just a week after the infamous team solidarity photo was tweeted, and the team returned to play, Coach Pinkel walked back his tweet and expressed his dilemma. On the one hand, he was a coach who stood by his players, which showed support and sympathy for his Black athletes and their concerns. On the other hand, his support seemingly endorsed a revolt directed at the system president and the university's inaction to improve Black students' conditions. Pinkel could not express that he felt something for Jonathan Butler, who was willing to die for change. He claimed the original tweet was not accurate, using the hashtag #CS1950 was a clerical mistake made by the staff person who did some of his tweeting (Palmer 2015). He said he was not supportive of the cause; he was supporting his players.

Read the code: Pinkel did what he believed was necessary to get "his" players back on the field, to get the upcoming game played. He did his part for the university. Ultimately, Coach Pinkel gave in to the racism, but his job was not saved; he announced his resignation shortly after. Pinkel paid the price for aligning with his players and being a part of questioning the school's racist institutional practices. Even though he was not fully aware of his actions' potential consequences, Pinkel was the third White casualty.

Many Whites take positions as potential overseers and master's wives, consciously, and unconsciously. Ben Trachtenberg, a White Associate Professor at the University of Missouri School of Law, talked about the challenge of diversity for Whites at large, predominantly and historically White institutions. "Public universities have a mission that makes them particularly susceptible to this tension. You have completely legitimate demands for greater racial inclusion and racial justice and the potential for backlash from taxpayers who fund the university." Trachtenberg presents a neutral analysis that was not neutral. The phrase "backlash from taxpayers who fund the university" is loaded with White privilege. The backlash is deemed legitimate and equal to racial inclusion and racial justice. He stressed "completely legitimate" and uses "demands" to speak about greater racial inclusion and racial justice but felt no need to discuss legitimacy or how taxpayers make unjust demands. Implicitly, one side is demanding and must be legitimized, whereas "taxpayers" are inherently legitimate and responsive rather than demanding. Deeper still, is naming "taxpayers" as a foil for legislators and the interests of White elites? Who is to say what the taxpayers want? Who are the taxpayers, and are they a homogenous group? Why would taxpayers oppose greater racial inclusion and racial justice? How do they understand inclusion and justice to be counter to their interests, and where did they learn that?

Racist Responders

Writing on Facebook, a former White Mizzou football player criticized the Black football players, calling their action "a pure lack of responsibility and ungratefulness. Playing football at the University of Missouri is a privilege, a privilege a lot (of) individuals would take from you if you are willing to give it up." He was Luke Lambert, a MU linebacker from 2007 to 2011. He continued, "Are you truly following through with the message sent to the public or will you still collect the monthly scholarship check to enjoy that night out, eat in the free dining hall, earn the free degree and enjoy the free gear handed to you during your tenure at the University of Missouri?" This White player could not comprehend the reasons why Black football players would protest. His perspective showcased a White proprietary claim to the university: Black players should be grateful for playing college football and attending Mizzou. In this way, Black athletes did not belong at the university; they had been given an exception to perform for Whites and earn profits for the university.

Tim Wolfe, MU system president, was the quintessential overseer. He was in charge and was to keep everything going smoothly. Instead, he allowed or was unable to prevent the situation from getting out of hand.

The day after Tim Wolfe stepped down, there was an anonymous, campus-wide death threat given to Black students via YikYak, a social media post; it read: "Some of you are alright. Don't go to campus tomorrow. I'm going to stand my ground tomorrow and shoot every Black person I see." Hunter Park, a nineteen-year-old college student (at Missouri University of Science and Technology), confessed to making the threats and risked being charged with a felony. He was arrested and released shortly after—potentially a fourth White casualty. Park lived in an affluent St. Louis neighborhood and had published posts that decried people's dislike of the wealthy (Burdziak 2016). He was given a suspended sentence and only five years' probation; his lawyer argued successfully that Hunter had no weapons or a car to drive to carry out the threat.

The Master's Act

In direct response to the football protest and within 30 days, two state representatives, Rick Brattin and Kurt Bahr sponsored a Missouri legislature bill to eliminate future boycotts from happening. The proposed law, House Bill 1743, states: *"Notwithstanding any other provision of law, any college athlete who calls, incites, supports, or participates in any strike or concerted refusal to play a scheduled game shall have his or her scholarship revoked."* Individual actions in resistance to the Black protest had failed. Brattin was set to use the weight of White proprietary claims in the state legislature. He described the protests as "completely horrific" and "anarchy" and added that the Black football players had behaved inappropriately (Sherman 2015). The proposed bill did more than outlaw similar actions; it proposed a final recourse: revoke an athlete's scholarship. He found fault with the players and their overseers, who had not done anything to protect Missouri. Brattin said that he knew that the bill would not pass—it would have been unconstitutional—and he and Bahr pulled their bill, but Brattin continued to express concern over disorder at the university.

> My bill was filed to generate discussion on what I believe is an extremely important topic, and one that deserves deliberate consideration. While I am withdrawing the legislation, I hope the conversation will continue so that we can take steps to ensure the University of Missouri is providing a stable, positive learning environment for our young people. I sincerely believe students should be able to express their viewpoints. Still, I also believe our flagship state university has to keep and maintain the order expected from such an esteemed educational institution ... We cannot have the student body, or in this case, the football team, going on strike and forcing out a school president. That cannot be allowed. (Perez 2015)

Brattin speaks about his bill with fire—he wants something to be done. The university is to provide a "stable, positive learning environment." But for whom? CS1950's mission was to represent what was missing and needed for Black students to have a stable, positive learning environment—and it was not against any other groups; it was pro-justice. Beginning with the events of 2014, the University had a full year to work and develop a plan to address Black students' concerns. But the "order" to remain unchanged was grounded in protecting Whiteness, White jobs, and White comfort. Brattin was clear: Black football players cannot in the future be allowed to determine the fate of the school's President.

A wider anti-Black student response was swift. Students, parents, alumni, and MU supporters—mounted a countermovement (Melchior 2016). They protested Black protesting. They invited ultra-conservative speakers to campus. They started clubs for "freedom of speech." White racist students bumped into Black students on "the strip," daring them to fight. They answered the call of anti-Black racism and were active in trying to keep Blacks in their place. A young White man entered a room where Black students were rehearsing a dance for homecoming—and—shouted racial epithets (Knott and Loufti 2015). MU's student body president, a gay Black male, was assailed by racial slurs from passengers in a passing pickup truck.

Students with old master's mindsets on campus emphasized the costs of racism; racism provided a false sense of superiority and loss of moral integrity (Kivel 1996). This led to some White students enduring affective costs of racism because they had/developed a reactive, irrational fear of Black students and other students of color that was not grounded in evidence but instead based on skin color (Spanierman and Clark 2021). Public spaces reinforced the race caste system. These White students conducted themselves with knowledge of the racial hierarchy; they yelled racial slurs without worry, threatening Blacks in public spaces. White empathy and guilt were low as these students had no cultural sensitivity and appeared to lack awareness of their White privilege—fifth White casualties. Regardless of a Black student's position—athlete, student body president, sorority/fraternity—they were not White and therefore were shunned, belittled, and less important. Racist White students knew that they were protected by the University police and by the University, that their words and defense would matter more than a Black student's. Black Lives and Black Students' Lives did not matter.

In the stratified White class structure, elite Whites manipulate White middle persons like the overseer and master's wife via racism. Wolfe sent an email letter on January 19, 2016, to members of the "Missouri 100," a group of prominent supporters, outlining the wrong thinking and shortsightedness of the athletic director, coach, and school chancellor (Wolfe 2016). He explained that better White leadership would have punished Black players rather than sacrificed the university's reputation and financial health.

Wolfe questioned why White men (the Chancellor, athletic director, and head coach) sided with Black football players when they should and could have stuck together to "teach a lesson." He used his business background and illustrated that the dollars did not add up—forfeiting the game was short-term thinking, losing control to Black football players was a greater threat to the system.

The Impact of Racism on Whites at Missouri

To preserve the balance of power in Congress between slave and free states, the Missouri Compromise was passed in 1820, admitting Missouri as a slave state and Maine as a free state. Except for Missouri, this law prohibited slavery in the Louisiana Territory north of the 36° 30′ latitude line. In 1854, the Missouri Compromise was repealed by the Kansas-Nebraska Act. Three years later, the Missouri Compromise was declared unconstitutional by the Supreme Court in the Dred Scott decision, which ruled that Congress did not have the authority to prohibit slavery in the territories.

Missouri was an exception—its latitude was northern but was a part of the Slave South. Missouri was a slave state *by choice*. Two hundred years later, we can identify remnants of that choice in the events that occurred before, during, and after the 2015 football protest at MU. The case of Missouri's football protest reveals how the university, its administrators, and students engaged in cultural, institutional, and individual racism to reproduce and reinforce White supremacy.

The different forms of racism are not just done to Blacks; they are done in the service of White elites, with costs to Whites middle persons—students, faculty, coaches, and administrators. What made these moments at MU a "big event" was a public discourse—a back and forth regarding right and wrong. There were a critique and denunciation of the Black students and athletes, and expressions from various Whites signifying that they saw Black actions as improper and a threat to the proper social order. In short, White middle persons were forced to express their concern and disgust with Blacks' denunciation of White superiority.

White supremacy conditions us to accept and feel racism as normal; Whites' desire to uphold normalcy shields the racial hierarchy and sense of group privilege. Whites threatened by other groups can see disruption as a bad thing, contrary to normal, and certainly not racist and problematic. Thus, they never questioned whether the way things are in the present is the way they should be. Even when it is in the White middle person's best interest to combat racism, many will not. Racism hides White class divisions and does not make gains for most Whites in the middle. The racism apparent in the "big event" and its backlash broke through the presumption that Whites are a monolith, are not stratified, and all gain.

During a racist "big event," people and organizations are stamped as heroes and villains, which has reputational and economic effects. The University became marked in two paradoxical ways. It was a place where liberalism and anarchy were ruling the day—Black students and athletes were able to get a White college president fired. It was also a racist place, so racially dysfunctional that there was a campus-wide death threat. The handling of Black protests by the University's administration and board of regents/trustees shows their investment in racism despite the potential damage inflicted individually and collectively on Whites involved—the university's president, chancellor, athletic director, football coach, student-athletes, and students. The reputational stench has created a loss in applications for admission and a crumbling of the flagship campus.

Conclusion

Racism is not inherent in Whites. Most Whites go along with what they are told should be their attitudes and behaviors toward Blacks and others. The White elites who dictate what to do exploit the majority as they use Blacks and racial differences to cover their interests. It is clear, as it was during slavery, that when a few Whites at the top have a personal agenda to profit (in this case, through a state university) they may effectively use White supremacy (racism) to cover this fact. They need racism to distract the White public and need their support to continue making profits in business deals, television rights, profits from concessions, commercials, and opportunities for tax write-downs. This is how a small minority of wealthy slave-owners remained wealthy and in power in a sea of White poverty over a century ago. They did so by exploiting slaves and keeping most Whites poor, and thinking they were superior to Blacks. This competitive stratification has been essential to racism: the plantation elite got nearly half a million poor Whites to fight and die in a Civil War on their behalf, a deadly impact of racism on White Americans.

The fear of a Black planet is a powerful tool that galvanizes and also hurts Whites. In the same historic Ted Koppel interview, Campanis spoke to Whites' fear of Blacks taking their position. "They've [Blacks] got everything," he said. "If they take over coaching like everybody wants them to, there's not going to be anything left for White people." This is precisely how Donald Trump used racism to fire-up his base and attempted to remain in office after he lost the 2020 election. Campanis articulated the White identity politics of the Trump political base and of the January 6, 2021, White "patriots" who assaulted the Capitol.

Racism is learned, reinforced in culture, institutions, and individual experiences. It is often subtle, covert, and beneath the surface, and other times obvious, violent, and deadly. We never entered a post-racial era because racism is undefeated. It "doesn't die, it multiplies," said the late Robin Harris

(1990). In 2020, America faced a formidable social movement that chanted "Black Lives Matter," and the in-the-streets protestors were overwhelmingly White. They sought to regain some of their humanity in calling for a change in policing. Some even asked for an end to racism. It remains to be seen if they can sustain their protest in this period after the Trump Presidency with so many other Whites who are openly racist and in opposition. Will anti-racist Whites be able to go beyond cultural and individual racism and genuinely seek to end institutionalized and systemic racism which will benefit some Whites?

Ending racism requires more than an acknowledgment of White privilege. Whites must be willing to acknowledge White fear to give up what they perceive as their power. They must identify and seek to root out their normal and everyday ways of thinking that lead to differences in resources, opportunities, and othering. This is the call for more Whites to be anti-racist. However, White Americans are not a monolith and are varied in their social location. Whites in the middle must accept what is perceived as White loss and do it while fighting an internal race war against fellow Whites who are not amenable to a change in the social order. At the heart of this, they must fight off the fear of a Black Planet, perpetuated and used against them at great costs to themselves.

Discussion Questions

Q1: What is the "fear of a Black planet?" How can it both galvanize and hurt Whites?

Q2: What happened to the White faculty and staff who supported the student-athletes' protest?

Q3: What happened to the university's enrollment and status when the university tried to quash the protest?

Q4: The protests at the University of Missouri happened before Trump became president. What implications does the "big event" have in the age of Trump?

References

Associated Press. 1999. "Schott's Reign of Terror Lasted 15 Years." *ESPN*. Accessed November 6. http://www.espn.com/mlb/news/1999/0915/59994.html.

Babb, Stephen. 2014. "A Brief History of Donald Sterling's Alleged Racism." http://bleacherreport.com/. Accessed July 26. https://bleacherreport.com/articles/2042877-a-brief-history-of-donald-sterlings-alleged-racism.

Bogage, J., and A. Reiss. 2015. "Football Team Says it's 'Ready to Play' After Wolfe's Resignation." *The Columbia Missourian*, November 9. https://www.columbiamissourian.com/sports/mizzou_football/football-team-says-it-s-ready-to-play-after-wolfe/article_317779ac-8704-11e5-aeee-2b5ec204484f.html.

Bonacich, E. 1973. "A Theory of Middleman Minorities." *American Sociological Review* 38: 583–94.

Bowser, Benjamin P. 2017. "Racism: Origin and Theory." *Journal of Black Studies* 48 (6): 572–90.

Brooks, S.N. 2016. "The Fire This Time: A Context for Understanding the Black Athlete Protests at Missouri." *Phi Kappa Phi FORUM* Summer: 18–21.

Burdziak, Alan. 2016. "Hunter Park, Who Posted Racist Threat During MU Unrest, Gets Probation." https://www.columbiatribune.com/, June 17. https://www.columbiatribune.com/973809c7-6f10-58a5-83ce-1961008e35df.html.

Comeaux, E. 2012. "Unmasking Athlete Microaggressions: Division I Student-Athletes' Engagement with Members of the Campus Community." *Journal of Intercollegiate Sport* 5 (2): 189–98.

Douglass, F. 1845. *Narrative of the Life of Frederick Douglass, an American Slave*. Boston: The Anti-Slavery Office, No. 25 Cornhill.

Harris, R. 1990. Robin Harris. In *One Night Stand*.

Jo, L. 2015. I am a White, Non-Racist, Non-Violent Mizzou Student and I Actively Oppose #ConcernedStudent1950, http://totalsororitymove.com/.

Kivel, P. 1996. *Uprooting Racism: How White People Can Work for Racial Justice*. Philadelphia: New Society Publishers.

Knott, K., and E. Loufti. 2015. "LBC Homecoming Royalty Harassed at Traditions Plaza." *The Maneater*, October 5. https://www.themaneater.com/stories/campus/lbc-homecoming-royalty-harassed-traditions-plaza.

Melchior, J.K. 2016. "New Revelations about Mizzou Crisis." *National Review*, April 26. https://www.nationalreview.com/2016/04/mizzou-firings-chronicle-campus-crisis/.

Palmer, T. 2015. "MU Coach Gary Pinkel Distances Himself From Activist Group in Radio Interview " *The Kansas City Star*. https://www.kansascity.com/sports/college/sec/university-of-missouri/article44174805.html.

Perez, A.J. 2015. "Missouri Legislator Withdraws Bill to Bar Student-Athletes From Protests." *USA Today*, December 16. https://www.usatoday.com/story/sports/college/2015/12/16/missouri-legislator-rick-brattin-withdraws-bill-bar-student-athlete-protests/77417410/.

Rhoden, W.C. 2006. *Forty Million-Dollar Slaves*. New York: Three Rivers Press.
Sherman, R. 2015. "Everything About That Missouri Bill to Ban College Athletes From Protesting Was Stupid." https://www.sbnation.com/, December 16. https://www.sbnation.com/2015/12/16/10214874/missouri-bill-athletes-protest-rick-brattin.
Spanierman, L., and D. Anthony Clark. 2021. "Taking White Racial Emotions Seriously: Revisiting the Costs of Racism to White Americans." In *Racism's Impacts on White Americans in the Era of Trump*, edited by Duke Austin and Benjamin Bowser. New York: Palgrave.
Sterling, D. 2019. "Transcript of Alleged Donald Sterling-V. Stiviano Recording." Last Modified February 25. https://www.thoughtco.com/alleged-donald-sterling-v-stiviano-recording-2834757.
Weinbaum, W. 2012. "The legacy of Al Campanis." March 29.
Wolfe, T. 2016. "Text of Tim Wolfe's Letter Criticizing University of Missouri leaders" *The Kansas City Star*, January 27. https://www.kansascity.com/news/local/article56868723.html.
Yan, H., and A.C. Stapleton. 2015. "University of Missouri Students Report Threats; Police Quell KKK Rumors." *CNN*, November 11.
Zimbalist, A. 1999. *Unpaid Professionals*. Princeton: Princeton University Press.

11

Education: From "All Children Can Learn" to "You People"

Patricia A. Maloney

Abstract Patricia A. Maloney shows the subtle impact of racism on White school teachers. White teachers, especially those in less supportive and resource-poor schools, lost their initial sense of empathy for their students and self-reflection. Five years into their careers, they had less job autonomy, felt more alienated from their work and students, reported less job satisfaction, experienced more burnout, and had higher turnover than White teachers in more supportive environments. Ironically, they begin their careers enthusiastic and empathic toward their minority students but ended up indifferent and even hostile five years later. They began believing that they were teachers and educators. Instead, they found themselves in institutions whose primary mission was to house and control students of color to replicate racial hierarchy. Education was an institutional pretense. The students and parents knew this and treated the teachers accordingly. The students were oppressed, but so were the White teachers.

Keywords Education · Teachers · Whites · Racism · Black students · Burn-out

P. A. Maloney (✉)
Texas Tech University, Lubbock, TX, USA
e-mail: patricia.maloney@ttu.edu

D. W. Austin and B. P. Bowser (eds.), *Impacts of Racism on White Americans In the Age of Trump*, https://doi.org/10.1007/978-3-030-75232-3_11

In America, teachers are still generally White, middle class, and female. This is true not only in suburban schools, but the urban ones mostly populated by lower-income students of color. This phenomenon—the racial and class mismatch between students and teachers—can have many effects on students. Rong (1996) posed cultural congruence theory, which argues that people of the same culture will have a more positive baseline view of each other than of people of other cultures. This theory accurately predicted that White teachers scored Black students more harshly than White students on identical behaviors. Why? What are the effects of this mismatch on teachers and students?

One way that we can examine why tension arises when members of different classes attempt to communicate is called cultural capital. This concept, developed by Pierre Bourdieu, has been utilized to understand how different classes have different values, behaviors, and frameworks for seeing the world. Formally, Bourdieu sees it as "a theoretical hypothesis which made it possible to explain the unequal scholastic achievement of children originating from the different social classes by relating academic success…to the distribution of cultural capital between the classes" (1986, 82). Cultural capital is class based: one's tastes and ways of seeing and valuing the world are based on what is valued in one's social class. This then affects how one values others with differing cultural capital.

One's cultural capital implicitly affects how one communicates with others, both in terms of content (word choice and what is communicated) and tone (the volume and level of formality and emotion). Thus, one's cultural capital can provide preferred rules for interaction (Lareau & Horvat 1999; Lareau 2003) that make a person feel uncomfortable if not followed. Therefore, a low-income student may speak to a teacher in a way that the student considers polite, yet the teacher, who could come from a different class, may construe it as disrespectful. At the very least, if the two individuals are unfamiliar with each other's cultural capital, they might still assume best intentions, but may never be fully comfortable. Such lack of comfort will have effects on the closeness of the relationship and the student's academic growth—it's hard to develop the necessary trust when mismatched. Levy et al. (1997) concurred, finding that Latino and Asian students "read" social situations and teacher authority differently than non-immigrant White teachers. This may indicate that there is a communicative chasm between students of color and White teachers. Carter (2003) correctly notes that many low-income students of color can code-switch between the cultural capital of the dominant class and their at-home cultural capital, but this requires extra work on the part of the student.

In linguistics, some research on this mismatch has concluded that a person's language is so inherent to identity that attempting to erase or change that language can affect the person's feelings of self-worth (Laneheart 1998). So, when teachers are trying to instruct students in knowledge and skills, they are also implicitly instructing them in the cultural capital and communicative style of the middle class, which may make students feel "othered" in the classroom.

How the Structural and Personal Meet

Thus, we have teachers who are mostly White and middle class with the accompanying cultural capital. Those teachers themselves attended mostly middle-class schools, but can work in very different ones. The teachers in this chapter are in exactly this position. They are White, generally born middle to upper class, and have had little experience communicating with people of color who have a different cultural capital. The larger study also included teachers of color, but, given the topic of this book, I focus here on the experiences of the White teachers.

To gather these data, I spent three years with these teachers that began as they left training programs that explained students' low academic achievement on structural factors like unequal school funding and neighborhood segregation. I observed their classrooms, interviewed them, and listened as their views about race changed. I am a White middle-class female former teacher, which affected my ability to build rapport: they felt that I understood their job stresses and were able to speak about race more frankly than if I had been a person of color.

Overall, I found that these White teachers were hugely impacted by their interactions with students. As can be seen in Fig. 11.1, the White teachers' ideas about the cause of poor academic performance in students of color changed significantly during observation. I identify two types of interactions that changed teachers: (i) micro-level, interpersonal interactions with students and parents who do not share their cultural capital/communication style and (ii) interactions with the school system and other peers/colleagues in the school system that formed teachers' definition of the best way to pursue student success. No single chapter could talk about all of the different ways that teachers interact with the school system, so I enlarge upon what I term two indicators of institutional racism: (a) predetermined curricula and textbooks and (b) the school district's often unattainable standardized test goals.

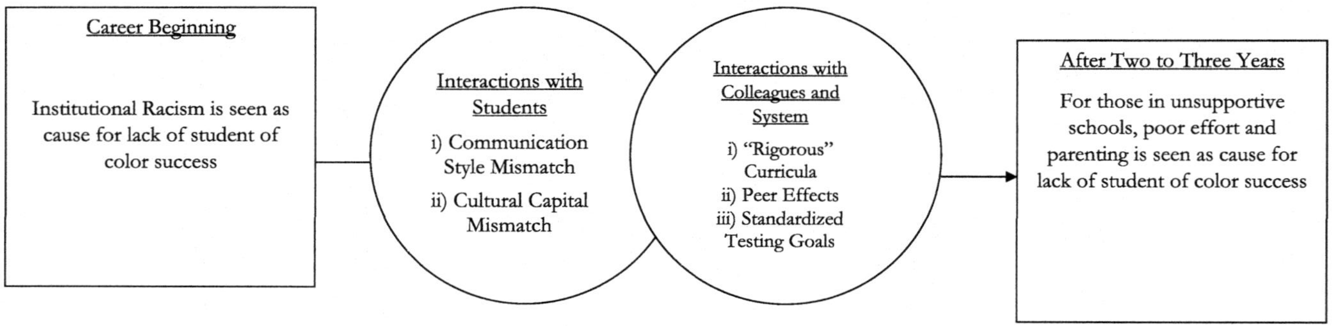

Fig. 11.1 How unsupported teachers change their views

These two types of interactions—with students and with colleagues/the system—are how I structure this chapter. After three years, instead of blaming external, structural shortcomings, some teachers blamed the students (and parents) for academic failure. This materially changed the content and tone of teacher/student interactions, causing more alienation and lack of growth, thus proving the teacher "right" about the students.

One large factor that affected the proportionality of the teacher's shift is the level of support s/he received in both instruction and classroom management from the administration, as well as the level of informal support s/he received from colleagues in learning how to teach effectively. By proportionality of the shift, I mean: every teacher in my sample began by espousing the belief that structural racism caused students of color to not be able to receive an equal education. By the end of data collection, every teacher had, at some point, expressed some variation of individual blame: that the students and/or their parents had something intrinsically wrong with their values causing the lack of academic success. However, the frequency and strength of that shift varied wildly.

Certainly, not every White teacher reached the same conclusion that students and parents were completely to blame. In higher resourced schools with supportive principals, many beginning White teachers mostly maintained their beliefs about structural racism despite a mismatch in cultural capital. In other schools, which tended to be low-support and -resource institutions, teachers were more likely to at least partially blame the students' poor outcomes on students' personal failings or parents' laziness or poor priorities. Generally, the age of the student affected the proportionality of blame—parents were blamed for younger students. Older students were considered to have enough personal agency to be able to "fix their own life," even in the face of parental neglect or abuse. This may be due to what Ferguson (2001) calls "adultification," or the tendency for Whites to see children of color as more agentic than their White peers.

Unsupported Teachers

First, we'll look at how this process occurred in schools that didn't support their teachers through either formal means, like instructional coaching, or informal means, like colleagues sharing information on how to most effectively interact with parents. In such schools, teachers form their own socioprofessional networks, mostly focused on stress relief and cracking jokes about students or administrators. Such schools are difficult for teachers to work in since they tend to be under-resourced and are particularly susceptible to the pressures of standardized tests. More specifically, they have students who are disproportionately behind academically, so curricula and tests that rely on previous knowledge are asking a great deal from teachers: to catch students up multiple grade levels to a score that is deemed "proficient" by the district.

The only time I saw a teacher making what could be read as a racially-based insult happened in such a school. It was, in his telling, not deliberate, but caused an uproar in the sixth-grade classroom that took three veteran teachers and a dean of discipline to quell. The (White) teacher had been telling off a group of students about their lack of work and constant talking. In doing so, he said: "You people better get it together or you're going not going to pass." The students immediately reacted to the phrase "you people" and said, "what do you mean, you people? Niggers? We're all stupid?" I was present that day and believe that this was accidental on the teacher's part. His face certainly blanched when he realized what he had said. However, could the use of "you people" have been unconscious racism? Certainly. Afterward, when we were discussing this incident, he expressed his anger with the students who responded to the "you people" as well as the rest of the class that had capitalized on the chaos:

> Why do they have to be so touchy? OK, I don't mean that - I get why their antenna would be up about racism, because I'm sure that my kids have experienced some horrific things. I have no idea what people who look like me have done to them. But, I'm in their school, trying to make things better for them. I'm not saying that they have to be grateful, but can't they realize that I'm trying? Shouldn't I get some credit and trust for that? Hah, maybe this is what they're talking about - I'm upset that I'm distrusted solely on my race, despite my actions to the contrary.

Here, this teacher is being introspective and self-critical about his feelings about race. He's angry with his students because he feels like he should be given credit for his efforts, but then catches himself and empathizes with them ("my kids"). This happened early in his first year of teaching. But, by the end of his second year, the empathy had been beaten out of him by the stresses of such a high-conflict school. Then, this same teacher said,

> There's something wrong with them. Why can't they just sit down, shut up, and listen? I could teach them a ton if they'd just shut up. They're never going to get anywhere if they can't behave. It's their parents, too - they come in with a chip on their shoulders and a sense of entitlement. Send your damn kid to

> school with a fucking pencil! I can't believe that those people would rather give their kids money for potato chips for breakfast than actually help them get out of this hellhole. I don't know what's wrong with all of them.

Even though he had previously experienced a memorable consequence for "you people," he uses the same generalization here, referring to his students' parents as "those people." His anger causes him to see them as not only a monolithic group, but an oppositional one: people whose priorities are so skewed that they would rather spend money on junk food than nourishing breakfasts or supplies. Their cultural capital is wrong.

For this teacher, two years in a struggling and unsupportive school had cost him his empathy and self-reflection. His world has narrowed—no longer does he have the ability to be charitable or nuanced toward his students and their parents. Rather, he's angry at their behavior, sense of entitlement, and skewed priorities. His students aren't individuals, with different traits and behaviors that vary personally. Instead, he refers to his students as "they" and "them"—they are a group that does nothing but enrage and insult him.

Without a doubt, this shift negatively impacts the students of color. Even though it was certainly not his intention to harm his students, by losing sight of the structural racism that the students face and becoming less empathetic to their circumstances, the White teachers became agents of the structural and institutional barriers that they previously decried. However, the White teachers are impacted, too. The price they pay is both personal (a sharp decline in mental and physical health) and professional (a decrease in job satisfaction and inability to invest more in their pedagogical practices). The White teachers thus become disillusioned and take an oppositional role with students and their parents, to varying extents. When this happens, all suffer.

Factors That Cause This Shift for Unsupported Teachers

There are several linked factors that shifted teachers from blaming a system biased against students to blaming the students/parents: curriculum that is ostensibly culturally responsive and rigorous, but actually ignores racism; interactions with students/parents who do not share the teacher's cultural capital; and interactions with peer teachers who encourage an "us versus them" mentality.

"Rigorous" and Racist Curricula at Unsupportive Schools

First, the curriculum: the teachers at unsupportive schools were basically given the keys to their classrooms, textbooks, curricula, and left alone for about six weeks at a time. This particular time schedule was based on standardized tests. Every six weeks, the students would take "benchmarks," or short multiple-choice tests about the material that they covered over the previous six weeks. Such tests were supposedly not cumulative, but did assume that the students had mastered all knowledge and skills from the beginning of the school year. For example, one teacher taught Algebra 1. Since she had taught the quadratic formula during the first module, students were asked to reuse it at the end of the year to answer a question about factoring.

As the teachers expected, scores on these tests and on the eventual cumulative end-of-the-year exams tended to be low. This caused higher levels of scrutiny on the part of administrators on these teachers, and

> …endless meetings and professional developments with whatever overpaid consultant. You know what we need? More teacher aides. Textbooks and curricula that are focused on student need and something they can actually learn from. Smaller classes. Students who aren't six years behind and can actually learn something would help!

This teacher said the above about halfway through her first school year, after a particularly low benchmark. The order of the list of elements that she reports needing to help educate her students mirrors her mindset about what (or who) is to blame: she starts with institutional issues like curricula and class size, but ends by noting that those factors are necessary, but not sufficient. She recognizes that the students are part of the equation, too, but even her words about them reflect this tension between the institutional and the individual: "students who aren't six years behind" versus "can actually learn something." Students being six years behind can be seen as the fault of the system—who allowed them to fall so far behind? But, the "can actually learn something" is more of an individual attribute. That is a flaw of the student, not the system. Later in the year, this same teacher expressed that

> You know what's sad? I had a little daydream while driving to work this morning about getting into a car accident, and how wonderful that would be. I'm not talking about something that would kill me. Just break an arm. Something that would be a legitimate excuse for not going into work so I can sleep and not fucking deal with the constant bullshit. I spent forever trying

> to teach them how to do basic rhetorical devices – seriously, like two weeks on what a damn metaphor is. They were all ready, we took the exams and they're idiots. What the hell? Basic questions that I had been drilling them on all week and all of a sudden, duuuh. So, now I've got [Principal] screaming at me because they couldn't fucking remember a damn thing and she's got us doing this stupid rap workbook thing to 'get their interest.' That's racist as hell. Just because they're Black, they're going to like rap? But, you know what the damndest thing is? They're into it. Why can't they learn the real stuff?

Beyond the fact that this teacher is so desperate for a break that she's fantasizing about a car accident, this quotation is full of her frustration at her students rather than at the system. She notes her work in trying to teach them metaphors and intense disappointment at later test scores that don't reflect this effort. The students are seen as stupid and not receptive to that knowledge, so she blames them for her principal "screaming at her"—obviously not a supportive supervisor interaction—and forcing her to implement a supplementary curriculum based on rap. This book, which she showed me a copy of, did actually contain accurate information about rhetorical devices alongside lyrics. However, it was also poorly edited and full of grammar and spelling errors, and contained coloring pages that encouraged high school students to color in rappers' clothing.

> When asked what "real stuff" was, this teacher gave me a withering look:
> No, it's not all old White men. Absolutely, Toni Morrison should be right next to Shakespeare. I just feel like coloring books are dumbing it down. It's the racism of low expectations, but...they're quiet. And calm. No one's throwing textbooks. Are they actually learning something with this that they might not with the real stuff, so maybe I'm actually part of the problem by wanting them to do the textbook? Am I giving in to this rap crap because it's just so much easier?

Given their preparatory coursework on structural racism, all the White teachers in this sample were familiar with what they referred to as the soft racism of low expectations. This is the concept that not holding their students of color to the same expectations as White students was seemingly being nice in the short-term, but harming the students in the long term. So, she acutely feels the tension of not wanting to fall prey to this. But, the lure of a quiet classroom with no behavioral issues while the students did schoolwork, even below-level schoolwork, is hard to resist.

Teachers reported that the content and perceived rigor of their curricula indirectly correlated with their students' behavior. That is, the more time they spent trying to encourage critical thinking and the school district's prescribed curriculum, the more behavioral problems they had. While there are certainly multiple reasons for this correlation, not least of which is that students can get anxious when asked to think abstractly, one teacher pointed to how her history text discussed race as a reason for students to rebel against the material and therefore her, as the agent of that curriculum:

> we're studying the civil rights era. I thought they'd be totally into it because I'm an idiot, so I got [another new White teacher] to work together with me on a unit where we read the Dream speech and discussed the historical implications of nonviolent protest and then she'd do the language arts stuff. I'm all bright-eyed and bushy-tailed because I think I'm being all cool and go, Black people, right? But this is like the fifth time they've had some White chick read them this speech on how they should be good little Black protestors and not hurt anyone…It's trite and insulting. They completely disengaged. I lost them for a month.

This history teacher followed the curriculum's recommendations for how to teach the civil rights era (the cross-disciplinary idea on Dr. King's speech) to find that it was tone-deaf. She, a White female, was seen as an agent of just another type of racism that attempted to assure her students that inequality was only in the past. Unsurprisingly, this was discordant with their lived experiences.

Awkward Interactions with Students

Beyond curricular factors, teachers at unsupported schools, who tended to stay alone in their classrooms when not chatting in the teachers' lounge, reported being uneasy in their interactions with students. The single most common sentence they said about their students was a volume complaint: "Why do they have to be so loud all the time?" This was often followed by a sentiment indicative of the insecurity that underlaid these interactions: "I can't tell when they're making fun of me," indicative of a communication and cultural capital mismatch.

That is, while the teachers and students certainly understood each others' words, many teachers reported not understanding the intent behind the students' communication:

> They know I'm White. They know that I don't always know what they mean, so they deliberately pretend to say hi to me in a loud way, but it's just making fun of me, especially when they're in a big group. But, I can't call them on it

because they're just saying hi. They don't respect me and know how to play the system.

This teacher directly connects her "Whiteness" with not knowing what the students intend in their interactions, particularly when the students are in a group and she's the only White person. She, like other teachers, implicitly doesn't trust them, which was betrayed not only in her curt responses but also in the slight fear she evinced when talking to them. Indeed, sometimes she ignored students when they did this—especially in the hallway, which she considered their space.

This difference in interaction rules, as mentioned in the section on cultural capital above, has definite implications for the closeness of the relationship between teachers and students. Students, particularly students of color, can tell when someone is uncomfortable. This has predictable effects on student/teacher interactions and, therefore, academic achievement. I spoke with some students to find out if this was just the teachers' perceptions. After building up a great deal of trust with a group of high schoolers, I asked them about a time when I had seen them loudly greeting a new (White) teacher:

Student 1: We just want to say hi! I want a good grade.
Student 2: For real, though, sometimes you mess with her that way.

These students are partially confirming what the teacher thought—that they sometimes "mess with her" in ways subtle enough to not be formally disciplined.

These cultural capital and communicative barriers caused by racism had an unsurprising outcome: teaching and mentoring of students suffered. The impact of racism on White teachers was apparent. They grew to fear and resent their students, causing them to not have effective lessons and suffer higher levels of occupation-related depression. In turn, students noticed and reacted to these negative emotions, hindering their learning and intellectual growth. This cycle was predictable, exponential, and tragic in its effects on both teachers and students.

You're with Us or Against Us

All the teachers pointed to their colleagues as key for "getting through the day." The teachers at the unsupportive schools mentioned that some of their colleagues of varying racial backgrounds, particularly older ones, were formative in shaping their views toward students of color. These teachers would mention the "good old days," when the school was demographically heterogeneous, to talk about the differences between the current students (who lacked respect and academic skills, in their eyes) and the previous student population that "had parents who at least cared!" One of the teachers mentioned that her dean had the acronym "GIGO" on his wall, to remind him that "garbage in means garbage out." Such attitudes were contagious and hard to resist—because these teachers had little support elsewhere, they clung to their colleagues and wanted to present a united front to the students.

The Supported Teachers and Factors That Affect Their Shift

For teachers at supportive schools, this shift was much less severe, if at all. These teachers had access to aides for a much-needed other adult in the room, instructional needs-based coaching, and content area coaches who could "translate" the district-provided curriculum for their students' needs. The lack of chronic, devastating stress meant that these teachers could see the "child in the student. He's not a bad kid, just reacting to a horrible home and school by misbehaving because he's angry. We call them KWLMs, or kids we love the most." This was from a second-year middle school teacher.

This theme was repeated by other teachers in other supportive schools: because they were receiving needed classroom help and did not have constant oppositional interactions with students, they were able to maintain their belief that situations make bad behavior, rather than bad children and parents make bad situations. This is not to say that these teachers didn't get frustrated with students—they did—or didn't become annoyed with parents. The difference is that these teachers were able to differentiate between individuals and the group. That is, just because this unreasonable parent is Hispanic does not mean that all Hispanics are unreasonable.

Of course, teachers at supportive schools were not immune to racism. As White teachers with predominantly Black or Hispanic student populations, they were certainly aware of race:

> Teacher: Of course I know I'm the gringa and all the parents are looking at me like I'm going to hurt their kids. It can be hard when I see that this kid is ignoring me in class because he heard his mom say that I don't know what I'm doing, but he listens to the Hispanic teacher who his mom loves. I know I'm the outsider here. I grew up in a totally White environment, so it's good to feel it from the other side. I'm learning Spanish from that other teacher. [The parent] is just going to have to get used to me.

This teacher is aware of race as a factor in her classroom, but approaches it with an understanding of her own agency: she can be a solution. A similar theme of understanding one's own place in perpetuating racism characterized another teacher's thoughts:

> So, I was coming out of a meeting with my [Black] coach, and it went really well, so I was on a high. This parent was in the hallway behind me, and I apparently was walking too slowly for her because she remarked really loudly that 'some people don't know how to get out of the way.' First, I thought, 'why do those people have to be so trashy?' But that's not right. It's not that Black people are trashy, it's that this woman is. I talk all day with lovely Black people.

Because this teacher had the emotional energy from a positive interaction with a colleague of color (and other positive interactions with students and parents), she was able to check her bias that all of "those people" are trashy in favor of understanding that just this particular parent who happened to be Black was a pain. This was a theme—teachers who had a "store" of positive interactions with students and colleagues were able to have more emotional resiliency and thus more flexible rules of interaction with others who weren't necessarily a part of their culture.

The more supportive schools are a marked contrast in regard to the impact of racism on White teachers than the less-supportive schools. The additional resources and positive interactions at supportive schools help to combat the structural racism that students face by empowering more empathetic and effective teachers, which reduces the racism that the students experience. At the same time, teachers in schools that do a better job combating structural racism report better mental and physical health, as well as tenure in the profession: fewer of them had left at the end of data collection.

An "Edited" Curriculum

While all the public schools in a district must use the same curriculum, teachers at more supportive, less structurally racist schools had more autonomy. They were able to "edit" the curriculum more. They felt less pressure about standardized tests from administrators, either because the school was already performing well, or because the administrators simply didn't care as much. So, they were able to move units around to fit their schedule or use different means to achieve the same learning outcomes. These were generally small edits, but the presence of staff like instructional coaches and less overall teacher turnover meant that the teachers could think through changes—luxuries that the unsupported teachers did not necessarily have, which led to the poorly received Dream speech unit described above.

Semi-Awkward Interactions with Students and Peer Effects

The White teachers at supportive schools reported the same initial difficulties of non-matching cultural capital with the students, but were able to overcome that through the support of their colleagues (particularly those of color) and a history of other positive interactions with students:

> Initially, I was nervous about joking around, not only because I'm a new teacher, but because I wasn't sure about the rhythm of the jokes. It seemed like some of the other teachers were saying things that were a little bit mean, but the students laughed and saw it as friendly. I was worried that if I did it, they would think I was really making fun of them. But, after watching for a bit and starting to do it with other teachers who are Black, I trusted myself more.

This teacher practiced new rules of interactions with friendly colleagues before trying them out on students. He was able to form a more comfortable relationship with students and build trust because he was supported pedagogically and socially by his colleagues, thus allowing both him and the other White teachers at such schools to not prioritize the rightness of their cultural capital out of insecurity and stress. These sorts of interactions allowed the students of color to more thoroughly learn with their White teachers because they help mitigate the effects of a cultural capital mismatch.

Conclusion

The White teachers in this sample entered their low-income and predominantly minority classrooms uniformly espousing the notion that structural or institutional racism was why students of color are less likely to succeed in the American educational system. Their ability and willingness to maintain that belief was strongly affected by their school—those at schools that created supportive academic and social environments were able to stretch their cultural capital and consequent rules of interactions and recognize the validity of someone else's. They were generally able to see their students as worthy individuals, despite interactions and curricula that occasionally tested their resolve and were able to resist cultural congruence theory.

For the new teachers who were isolated and unsupported, the everyday and intense stresses of teaching in low-income classrooms with students who they saw as antagonistic reified the institutional racism inherent within the school system. These well-meaning teachers lost their ability to maintain a meta-level awareness of their local social situations and instead reacted to oppositional interactions with anger and hurt. As a consequence, they assigned negative qualities to those who they saw as hurting them, thus seeing their students of color and their parents as intrinsically flawed, rather than the more removed system that was causing these interactions. This impeded an intimate and supportive relationship with students, thus hampering the conveyance of knowledge and skills.

The impacts of structural and institutional racism on students of color are well documented. Here, we see how racism impacts White teachers as well. White teachers, especially those in less supportive and well-networked schools, lose their sense of empathy and self-reflection, have less job autonomy, feel more alienated from their work and students, report less job satisfaction, experience more burnout, and have a higher turnover. Despite their initial desire to ameliorate institutional racism, they end up perpetuating it in unforeseen and clandestine ways. In more supportive schools, teachers and administrators do a better job at combating individual and institutional racism, and consequently have more meaningful and effective connections with their students, resulting in more positive work and learning experiences.

Schools thus become a prime example of the interplay of cultural/interpersonal and institutional racism. The conditions of the school system—caused by centuries of funding and infrastructure decisions that disadvantage low-income students of color—can be ameliorated by schools that are supportive of teachers, but not erased by interpersonal, micro-level actions by administrators and teachers. Tragically, those conditions can confirm and foster even greater levels of interpersonal racism, by "proving" to White teachers that students of color simply have something wrong with their cultural capital and themselves. Until these systemic inequalities are corrected, we can expect these effects on White teachers and their students of color to continue.

Racism in White Teachers During the Trump Presidency (2016–2020)

The three-year ethnography I refer to above took place before 2016, but I have continued to enter schools since then and speak with teachers in my continuing research about how racist and classist structures affect teachers and students alike. If I had to use one word to describe data collection on teachers in this time period, it would be "conflict." That is, teachers and schools are not separate from society in some sort of neutral ivory tower. Schools are intensely political spaces, both because they are affected by the literal policies of the Department of Education (run by Betsy DeVos during the Trump presidency) and because the human agents within them have their own thoughts, opinions, and ideas about the country's direction.

During this time period, I happened to be collecting data with my students in a more rural area of the country to add to our database on urban and suburban teachers. The contrast (or "conflict") between these more rural teachers and the urban ones with whom I started was startling. Almost to a person, the urban White teachers detested both Mr. Trump and Mrs. DeVos because of what they saw as rampant anti-intellectualism and a lack of investment in social services. Mrs. DeVos was generally decried because of her lack of experience in the public school system and removal of Obama-era rules that protected students of color and transgender students from disproportionate discipline. In this, urban White teachers and their students of color were generally in agreement, regardless of school type: Mr. Trump and Mrs. DeVos were a common enemy to be endured. The conflict was external: between them and Trump/DeVos. Sometimes, this was verbally spelled out in the classroom, particularly in History or English classes, when students and teachers would compare historical events like the Holocaust concentration camps to the immigrant internment camps on the U.S. border.

That was not necessarily the case for the rural teachers, who were also disproportionately White and female, but also often came from the rural area itself, or a similar setting. In America, teachers are disproportionately Democrats, but liberals usually cluster in urban areas. The rural area teachers we spoke with had a great deal more political conflict to navigate than their urban peers. They were a politically heterogeneous group themselves. Many reported feeling unsure about how to bring up current political events in the classroom or how to engage in difficult dialogues with students of varying class and racial backgrounds. The unity with students that the urban teachers took for granted was not something that the rural teachers felt, so they generally either declined to bring what they termed as "politics" into the classroom, saying that their classroom was based on content knowledge and skills rather than opinion, or sometimes shared their own more conservative opinions with which they felt their students would agree. I wonder, however, if that lack of an overt conversation about how politics affects their students is actually a political stance in itself? Is it the institutional privilege of the White teacher to not engage with the conflicts that such discussions would cause?

While my research on this is ongoing, I think it is. Indeed, the moral quandary that these teachers face is unenviable. Their desire is to lead their students to greater knowledge and skills—for their students to "see the truth for themselves." The issue is, whose truth? Current news reports in January 2021 are full of teachers who are telling their students that the 2020 election was stolen from Mr. Trump. While I certainly didn't find that a majority of teachers are saying this, particularly in a classroom setting, the concept of "alternative facts" and its accompanying conspiracy theories is a bell that is not easily unrung. Teachers are people, too, and no one is the villain in their own story. As people in a political society, teachers (of any race) are influenced by culture and it would be foolish to assume that influence somehow magically drops away at the classroom door. Are the White teachers less passionate when teaching about the civil rights movement? Did that English teacher shy away from the book by Elie Weisel because she didn't want to discuss the parallel between 1930s Germany and twenty-first-century America? It is impossible to know for certain, but the teachers generally report that they perceive their rural students to be more outspoken about their support for Mr. Trump than for other conservative politicians. Mr. Trump, a figure of great conflict, is also certainly a figure of influence that did not end upon Mr. Biden's inauguration.

Discussion Questions

Q1: This chapter focuses on the changes that occur in the minds and, therefore, the behaviors of White teachers in low-income, predominantly minority schools. What are those changes? What causes them?

Q2: What are the effects of those changes, both on teachers and students?

Q3: How does the concept of cultural capital play into how teachers and students interact with one another? Are there ways by which those cultural miscommunications can be corrected? Whose responsibility should it be to communicate for the best academic outcome for the students?

References

Bourdieu, Pierre. 1986. "The Forms of Capital." In *Handbook of Theory and Research for the Sociology of Education*, edited by J. Richardson, 46–58. New York: Greenwood.

Carter Prudence. 2003. "'Black' Cultural Capital, Status Positioning, and Schooling Conflicts for Low-Income African American Youth." *Social Problems* 50 (1): 136–55.

Ferguson, Ann Arnett. 2001. *Bad Boys: Public Schools in the Making of Black Masculinity*. Ann Arbor: University of Michigan Press.

Howard, T. 2010. "Culturally Relevant Pedagogy: Ingredients for Critical Teacher Reflection." *Theory Into Practice* 42 (3): 195–202.

Lareau, Annette, and Erin Horvat. 1999. "Moments of Social Inclusion and Exclusion: Race, Class, and Cultural Capital in Family-School Relationships." *Sociology of Education* 72 (1): 37–53.

Lareau, Annette. 2003. *Unequal Childhoods*. Oakland, CA: University of California Press.

Laneheart. 1998. "African American Vernacular English and Education: The Dynamics of Pedagogy, Ideology, and Identity." *Journal of English Linguistics* 26 (2): 122–36.

Levy, Jack, Theo Wubbels, Mieke Brekelmans, and Barbara Morganfield. 1997. "Language and Cultural Factors in Students' Perceptions of Teacher Communication Style." *International Journal of Intercultural Relations* 21 (1): 29–56.

Milner, H. Richard. 2006. "Preservice Teachers' Learning About Cultural and Racial Diversity: Implications for Urban Education." *Urban Education* 41 (4): 343–75.

Rong, X. 1996. "Effects of Race and Gender on Teachers' Perception of the Social Behavior of Elementary Students." *Urban Education* 31 (3): 261–90.

Solomon, R. Patrick, John Portelli, Beverly-Jean Daniel, and Arlene Campbell. 2005. "The Discourse of Denial: How White Teacher Candidates Construct Race, Racism, and 'White Privilege.'" *Race, Ethnicity, and Education* 8 (2): 147–69.

Part II

The Benefits of Anti-Racism

12

Social Movements: White Responses to Racist and Anti-Racist Movements

Pamela Oliver

Abstract Pamela Oliver provides a big picture to understand White American's unprecedented involvement in recent #Black Lives Matter demonstrations. She shows that the struggle for racial equality in the United States has never been Black versus White or a straight line of victories or defeats. A careful look reveals that Whites have been on both sides of the struggle against slavery and Jim Crow, and racism today. They have never been united in oppressing people of color or in enjoying White racial privilege. In reviewing the history of social movements, we should not be surprised that there is active White support for racial justice today. What is unprecedented is the scale of that support. This realization has profound implications for how we view social change and the likelihood that a reduction, if not elimination of racial inequality is possible.

Keywords Social movements · Whites · Racial inequality · Abolition · Jim Crow · Black Lives Matter

This chapter was written in the summer of 2020 amidst weeks of widespread protests after Minneapolis police officer Derek Chauvin killed George Floyd. Chauvin calmly knelt on Floyd's neck for nearly 9 minutes while Floyd pled

P. Oliver (✉)
University of Wisconsin Madison, Madison, WI, USA
e-mail: pamela.oliver@wisc.edu

D. W. Austin and B. P. Bowser (eds.), *Impacts of Racism on White Americans In the Age of Trump*, https://doi.org/10.1007/978-3-030-75232-3_12

for his life and called for his mother before losing consciousness. White people had different responses to the protests: many saw them as showing the need to do something about racism and police violence. In contrast, others saw the protests themselves as the problem.

These protests were part of the ongoing struggle in the United States over racism. They followed twelve years of heightened awareness of racial issues, including Barack Obama's election, the first Black president of the United States. There was the rise of anti-Black and anti-Muslim rhetoric in political discussions of President Obama's policies, and a wave of protests against police violence from 2014 through 2016, and the slogan Black Lives Matter. The campaign and then presidency of Donald Trump brought overt anti-immigrant policies and tacit support for White nationalism and their armed protests meant to intimidate others.

Racism is a structure of domination, a system of power relations organized around race that limits everybody's choices. This structure is re-created continually in movement–countermovement struggles between racist and anti-racist groups. Victories by one side are overturned by action on the other side. There has generally been more power on the side of White dominance, but this domination has never been complete or unchallenged. Political groups and elites intentionally promote policies they believe will benefit groups defined by race, ethnicity, economic status, gender, religion, region, or other characteristics. Individual Whites choose which of these policies or actors to support. Even though White people vary tremendously in their specific conditions and attitudes, they are the largest voting block. This means political elites have typically given their general interests and opinions great weight.

Some people see racial progress as a zero-sum game where every advance for Black people or immigrants is a loss for White people. Some people believe that anti-Black racism provides direct benefits to all White people. Others believe that the current system is either entirely racially equal or favors minorities. Some people believe that racism creates divisions among people that lead to harm for most White people. This chapter describes what groups have said in the past and mean in the present about the effects of racism on Whites as they struggled over policy.

A Brief Historical Recap

U.S. racial history is complicated and intertwined with different groups' economic and partisan political interests. Briefly, the United States was founded as a nation of European/White settlers who battled and displaced Indigenous Americans. Some of these White settlers owned African slaves or participated in the slave trade. Other White settlers were indirectly affected by the slave economy, some benefitting from it and others harmed. Creating a White-majority settler nation involved encouraging immigration from Europe and discouraging or banning immigration from other regions.

After the Civil War 1861–1865 over slavery, Northern White political elites made significant legal changes during Black Reconstruction, 1866–1877, with key Constitutional amendments that abolished slavery (13th), gave citizenship regardless of race (14th), and provided voting rights for men irrespective of race (15th).[1] However, the 14th and 15th amendments were negated in practice after Reconstruction by U.S. Supreme Court decisions.[2] These reversals affected White people as well. Social and residential segregation and employment discrimination were permitted that generally benefitted White people. Whites were able to maintain a sense of psychological or cultural superiority and reduced competition for jobs and other resources. However, political elites also used the same laws and practices to disenfranchise lower-income White immigrants in the North and Blacks in the South (Burnham 1965; Kelley et al. 1967; Piven 1988).

Black responses to White domination were ideologically diverse. There were the gradualist reformer, Booker T. Washington; the campaign against lynching by Ida B. Wells (Giddings 2008) and the NAACP (Francis 2014); and the Black nationalist movement of Marcus Garvey's United Negro Improvement Association. There were the Black women's club movement and the National Council of Negro Women (Giddings 1984; Upton and Bell 2017; White 1999); Black unionists, socialists, and communists (Heideman 2019; Kwoba 2017; Paul 2018; Perry 2009; Stillerman 2020); and many Black churches that supported Black advancement. Black activists had relatively few Whites allies in this period. Most were associated with the NAACP or the racially integrated Socialist and Communist movements.

[1] Indigenous Americans were generally classified as separate nations. Although some were made U.S. citizens by treaties, Indigenous Americans were not categorically defined as U.S. citizens until the United States passed the Indian Citizenship Act of 1924.

[2] The court permitted poll taxes, literacy tests, complex voter registration systems, and all White primaries to disenfranchise people. It ruled that citizenship was very narrow and ruled racial segregation as legal.

In contrast, the most visible face of violent White supremacy and the suppression of Black people in the late 1800s was the Ku Klux Klan (KKK), which resurged twice, first in the 1920s and again in the 1950s and 1960s (Blee 1991; Cunningham 2012; Dobratz and Shanks Meile 1997). Before 1960, many White politicians and elites, especially in the South, advocated White racial superiority, and most of their constituents believed them. The main branch of the White labor union movement prohibited Black or Asian members (Stillerman 2020). Although the women's movement of the 1840s was founded by abolitionists opposed to slavery and included both Black and White women, much of the White movement excluded Black women. By 1900, the main figures in the women's were responsible for this exclusion—Susan B. Anthony, Elizabeth Cady Stanton, Alice Paul, and the National American Woman Suffrage Association. They used anti-Black rhetoric in advocating votes for White women (Giddings 1984; Jones 2020).

Racism and Antiracism in and After the Civil Rights Era

Contrary to myth, the 1960s Black Civil Rights Movement operated in both the North and South and was opposed by most White Americans (Theoharis and Woodard 2005, 2003; Theoharis 2018). Before his assassination, Dr. Martin Luther King, Jr. was also seen as radical and dangerous by the FBI. Although Civil Rights workers associated with King protested nonviolently, they were met with violence in the North and South by police and White crowds. Some Black protesters fought back against attacks. Civil Rights gains in the South did not address racial economic inequality nationally. There were more than 160 Black urban insurrections or riots between 1963 and 1969, most beginning after police violence (Kerner et al. 1968). A majority (52%) of Black people saw "police brutality" as the major spark for these riots (versus only 13% of Whites), and 63% of all people agreed "until there is justice for minorities there will not be law and order" (cited in Taylor 2016, Chapter 4). There were also massive protests against the War in Vietnam and battles between police and mostly White anti-war protesters. The era seemed violent and even revolutionary to many people. By 1968, most Whites saw racism as a problem, with 53% in one poll saying they would support a 10% increase in their taxes to improve conditions for Black people (Louis Harris and Associates Inc. 1968; Urban America and Urban Coalition 1969). White people saw racism as a problem and were fearful of Black militancy.

In response to the urban riots and militancy, the Johnson administration and the White public supported poverty programs for a short period that benefitted White and Black people.[3] At the same time, King and other Civil Rights leaders long supported racially integrated labor unions and social programs to help working people of all races. In fact, King was assassinated while organizing a Poor People's Campaign for economic justice for all races (Theoharis 2018). With the assassination, the transition of the Civil Rights Movement to including economic rights was stalled.

The loss of King was traumatic. Many Black people shifted their beliefs away from the integration and nonviolence and called for Black self-defense. Research showed that Black understandings of Black Power emphasized racial equality and unity among Blacks, while Whites associated the term with Blacks ruling Whites or riots (Aberbach and Walker 1970). The FBI and large city law enforcement agencies worked to cause division within the Black Panther Party and other militant Black Power groups. The tactics they used were surveillance, disinformation, infiltration, and assassination of leaders (Bloom and Martin 2013; Davenport 2010, 2012, 2015; Dello Buono 1992; Goldstein 2001; Marx 1970, 1974). By contrast, the FBI and local police generally saw White supremacist groups as inherently harmless (Cunningham 2003, 2009).

Repressive surveillance and entrapment used against militants, after 1970, were turned to general crime control and surveillance in Black communities (Marx 1980, 1981, 1982). The research found that escalated policing, focused in Black areas (Carroll and Jackson 1982; Jackson and Carroll 1981), led to militarized policing and mass incarceration. Both trends negatively impacted everyone, including Whites (Oliver 2008). After the violent repression of Black militants between 1969 and 1972, the Black Power movement shifted from confrontation toward winning political offices and pursuing social and economic objectives through politics. A series of 1970s Black political conventions (Johnson 2007) sought to create a unified Black political agenda. Afterward, there was a steep increase in the number of Black elected politicians. Many political activists carried Black Power ideology into government and the professions (Bell 2014) to change the system from within.

[3] Although a higher percentage of Black than White people were poor, there were and are more poor White people than Black people because there are so many more White people.

Racism and Immigration

Immigration policy has contributed to structural racism as well. Briefly, the United States officially encouraged White European immigrants in the 1800s as settlers to displace Indigenous people from their lands and, after 1865, to replace formerly enslaved Black people as workers. Between 1860 and 1920, over 12% of the U.S. population were immigrants, nearly all from Europe (Migration Policy Institute 2020). Led by nativist elites and White workers fearing competition, White sentiment against first Asians and then Southern and Eastern Europeans led to a preference for Northern Europeans and sharply restricted immigration. These sentiments culminated in major changes in U.S. immigration law in 1920 and 1924. After the United States took control of the territory that had been Northern Mexico in 1848, Mexican citizens moved freely back and forth across the border as temporary workers whose primary homes were in Mexico. By 1970, immigrants were only 5% of the U.S. population.

The 1965 immigration law eliminated overt racial preference and substituted family ties and occupation as criteria for entry into the United States. Although the legislators intended the new rules to maintain White preferences with nonracial language, the new rules actually permitted a gradual increase in non-White immigration, especially of well-educated people from Asia. The 1965 law also imposed the first quota on Mexican immigration and created the problem of "illegal" immigration from Mexico by restricting what had been unrestricted migration of workers back and forth across the border.[4] US-born White people had high birth rates in the 1950s and 1960s but their birth rates dropped from the 1970s on. As a result, immigrants became a growing share of the U.S. population, reaching nearly 14% by 2019 (Migration Policy Institute 2020).[5] The proportion of non-Hispanic Whites declined as immigrants and their children were disproportionately Hispanic and Asian. Much anti-immigrant sentiment is fueled by fears of losing an Anglo-White majority in the United States. There are debates about the validity of these fears because at least half of Hispanics identify as White, and virtually all children of immigrants become English-dominant. Still, the

---

[4] The new rule capped permanent resident visas or "green cards" from any one country at 7% of total immigrants. Originally the rule limited only Mexicans and was intended to preserve Anglo-White cultural dominance. The rule now also caps immigration from the Philippines, China, and India. A series of economic crises in Mexico and Central America caused by international policies led by economic elites led more Mexicans and Central Americans to seek work in the United States between the 1970s and 1990s.

[5] For overviews of US immigration law across history see https://www.infoplease.com/us/immigration-legislation or https://www.fairus.org/legislation/reports-and-analysis/history-of-us-immigration-laws.

racial-ethnic mix of the U.S. population is doubtlessly changing. In addition to cultural fears, White workers both historically and today often fear that competition from immigrant workers drives down White workers' wages.

Racism, Political Alignments, and Conservative Political Movements

The two major political parties realigned around race after 1968. The Republican Party had been the pro-Black party in the 1800s. However, neither party supported Black rights between the 1890s and the 1930s. Between the 1930s and 1960s, both parties sought to attract both White and Northern Black voters (most Southern Blacks could not vote). The Democratic President Johnson and the Democratic-dominated Congress passed Civil Rights legislation in the 1960s, gaining Black support. The Republican party shifted to the "Southern strategy" of using fear nationally to attract White voters, who feared Black people and racial integration.

Politicians are generally dependent on wealthy donors and often advance and support economic and social policies that favor their elite sponsors. Republicans also sought to gain an electoral majority by using racist controlling images (see Dunham chapter), discouraging voters who might vote against them, and promoting fear of minorities among Whites. In response to social crises and popular movements in the 1930s Great Depression, Democratic presidents and legislators passed laws promoting union rights, worker safety, and social welfare programs like Social Security (1935) and "New Deal" anti-poverty programs (1964). These were paid for by high marginal tax rates that redistributed wealth from the prosperous to the poor. By the 1970s, strong environmental movements added to the list of reforms clean water and air through environmental regulations. However, by 1981 and the Reagan administration, elites businessmen such as the Koch brothers sought to roll those programs back, reduce taxes on the wealthy, and create regulations that would favor businesses over workers (Mayer 2017). Hidden elite interests successfully persuade Whites discontented with the Democratic Party to champion policies that favored big business through the Republican Party. The Republican Party gained political power repeatedly between the 1960s and the 2010s through coalition-building between Southern and rural libertarian interests.

Historically, faith communities were not generally partisan. After 1970, this changed. The Equal Rights for Women Amendment to the Constitution (ERA) passed Congress with overwhelming bipartisan support on March 22, 1972 and the Supreme Court ruled abortion as legal in 1973. The

Catholic Church had long opposed abortion and some Catholics immediately organized in opposition to the abortion ruling. Religious conservatives often believed in traditional gender roles and male dominance. Some White Southern Evangelical Protestants believed that God requires racial separation and has cursed Black people. Republican political elites intentionally adjusted their ideology to attract these White religious conservatives. As one part of the coalition, a group of Southern White Evangelicals originally formed to defend segregated White schools from losing their tax exemptions shifted their emphasis to being anti-abortion to draw in more Northern supporters (Balmer 2014). Similarly, economic conservative Phyllis Schlafly and other Republicans intentionally mobilized religious conservatives to oppose the ERA (Mansbridge 1986).

After 1968, White elite rhetorical resistance to racial change quickly shifted away from the overt defense of White supremacy. They claimed that the Civil Rights Movement had won, that legal equality was now ensured, that any remaining racial inequality was due to Black people's failures, and that any continuing attempts to enforce racial equality were unfair to Whites (Robnett 2020). Few White people living in the 1970s were willing to admit to ever having benefitted from the racial discrimination that had been legal until the mid-1960s. Polls showed that most White people endorsed the general principle of racial equality but opposed any concrete actions that would promote equality, such as school or residential integration, aggressive hiring or enrollment of minorities, or penalties for racial discrimination (Robnett 2020; Theoharis 2018). Another way to hide overt defense of White supremacy was to oppose "Affirmative Action," first used in President Kennedy's 1961 executive order. Schools, businesses, and government agencies were asked to promote racial equality in their organizations and to halt racial discrimination (See Pincus chapter).

Overt White Racist Groups

While organizations that wished to appeal to a broad base of Americans, such as the Republican Party, were covert in their racism, others who wished to appeal only to Whites were not. Some groups overtly promote White racial domination. Although they are a minority of White people in the United States today, they actively seek to influence White opinion. The Ku Klux Klan (KKK) was founded as an overt White supremacist group at the end of the U.S. Civil War. Its principal activities involved lynching and other acts of terrorism in the South and propagating White supremacy ideology. The KKK almost disappeared in the late 1800s but then rose again in the

1910s, reaching its peak in the 1920s. The wildly popular 1915 film *Birth of a Nation* fueled anti-Black hatred and extolled the KKK. As the KKK expanded to four million members, its agenda dehumanized Jews, immigrants, and Catholics as well as Blacks. It also emphasized Protestant Christianity and personal morality and opposed alcohol consumption. In both the Republican and Democratic Party conventions of 1924 anti-KKK resolutions were introduced and voted down.

Overt White supremacist groups like the KKK actively fought against Black rights through the 1960s, but they lost support after the shift to color-blind rhetoric in the 1970s. David Duke, the former Grand Wizard of the Knights of the Ku Klux Klan, resigned in 1980, saying that the KKK label was too stigmatized. He founded the National Association for the Advancement of White People and developed the ideology that Whites were now the discriminated-against group (Berbrier 1998a, 2002). KKK groups declined while other White nationalist or separatist groups, including Skinheads, Neo-Nazis, and White Christian Identity groups took their place. For a while they adjusted their rhetoric to make themselves more appealing to White conservatives who did not self-identify as racists, emphasizing themes of White cultural heritage and Whites as victims of discrimination (Berbrier 1998a, b, 2000).

White supremacists adopted the Internet early (Burris et al. 2000). As the Internet and social media technologies have evolved, White supremacists have used them to recruit new members and promote their ideologies. Online White racist attacks on people visibly advocating for Black or women's or queer issues have become common, along with "doxxing," the practice of publicly revealing people's home addresses and contact information and then subjecting them to harassment and death threats.

In the 1990s and 2000s, these groups avoided most overt collective action or protest, but they believed in violence (Dobratz and Waldner 2012). Individuals committed "lone wolf" acts of violence against minorities that appeared to outsiders as isolated but were encouraged by other group members (Blee 2005, 2017). Throughout the 1990s and 2000s, small groups of KKK or Neo-Nazis held rallies to draw attention and provoke responses. They were typically met by a much larger group of counterprotesters (Oliver et al. 2019).

From their Civil War and Klan beginnings, White racist groups have been willing to violently oppose the federal and sometimes state governments. In 1995, two White members of the militant Patriot movement, Timothy McVeigh and Terry Nichols, bombed a federal building in Oklahoma City, killing 168 people, including 19 children in a daycare center. They injured more than 500. This was the deadliest terrorist attack against White Americans before 2001. This event and other violent standoffs between armed White far-right militants and government officials led law enforcement agencies to do more surveillance of White right-wing and racial separatist groups. Consequently, White separatist groups consciously chose to lower their public profile, become more de-centralized, and avoid surveillance (Simi and Futrell 2020). They built safe spaces for themselves by homeschooling their children, emphasizing Aryan culture, and using cultural and social events and White identity music to recruit new members. White supremacist and nationalist groups are not closely connected to traditional conservatism (Berlet and Vysotsky 2006; Blee and Creasap 2010; Dobratz and Waldner 2006).

Black Movements 1980–2000

Black organizations struggled in the unfavorable political environment of the 1980s. Much Black activism was focused outward on the anti-apartheid struggle in South Africa. Jesse Jackson ran for president in 1984 and 1988 and organized Rainbow Coalition campaigns, which sought to create multiracial coalitions around maintaining and expanding social programs (Pierce 1988, 2003) but did not gain the Democratic Party nomination.

Discussions of race and policing in America returned to center stage in the 1990s. In 1991, a bystander video-recorded four White Los Angeles police officers brutally beating Rodney King. The video was shown on national television and widely discussed. The police officers were found not guilty in a 1992 trial moved to a White suburb. After the verdict, there were four days of multiracial rioting in Los Angeles with spin-off riots and protests in many other cities. In 1994, Black football star and media icon O.J. Simpson was accused of the brutal murder of his White ex-wife and her boyfriend; the case and the trial in 1995 attracted major media attention and racially polarized public opinion. In New York, a series of violent incidents by police led to a sustained waves of protests from 1997 through 2000 (Oliver et al. 2019). There were four days of riotous confrontations and months of protests in 2001 after a police officer shot a man in the back in Cincinnati. Black and non-Black organizations documented and then protested against "racial profiling" for "driving while Black" in traffic stops and other police abuses (Browne 1999; Camp and Heatherton 2016). A survey in 1999 found that 59% of Americans thought the police engaged in racial profiling, and 81% thought it was wrong (Cited in Taylor 2016). Several new Black-led organizations were formed in the late 1990s to work for criminal justice reform.

One was Critical Resistance, which opposed the "prison industrial complex" (Critical Resistance 2018). Another was Incite!, which addressed both official and personal violence against women (Ransby 2015; Richie 2012). The Radical Black Congress (RBC) brought together activists from a broad range of ideologies (Brewer 2003; Fletcher and Rogers 2014; Ransby 2018) and emphasized education and the criminal legal system. The RBC folded in 2008 but Critical Resistance and Incite! are still active parts of the current Black Lives movement.

Also important was the 1995 call by Louis Farrakhan of the Nation of Islam (Black Muslims) for a Million Man March in Washington, DC, for Black men to repent and take responsibility for their families and communities. Up to a million Black men attended the rally. There were spin-off and follow-up rallies and a general rise in Black civic activism, although White people seemed to pay little attention to this.

In the mid-1990s, many White people supported the passage of anti-immigrant laws that removed the right to make "welfare" payments to legal immigrants and required English-only instruction in schools, while pro-immigrant groups protested in response.

Many Americans, White and Black, discussed and debated the issue of slavery reparations after the January 2001 publication of Randall Robinson's *The Debt*, which advocated reparations. White conservative David Horowitz received publicity in 2001 by demanding that campus newspapers publish his inflammatory editorial that ridiculed reparations arguments and provoked Black and other students' protests.

2000 to 2020

The escalation of Black protest and activism around criminal justice and reparations was cut short by the 9/11 terror attacks. Attention turned to the War on Terror and then the Iraq War and anti-War protests (Oliver et al. 2019). Muslims were the targets of both law enforcement and racists. New issues arose for Black activism after 2005. The failure of the Federal government to respond to Hurricane Katrina in New Orleans was the most prominent example of racism. There were mostly peaceful but disruptive protests in New York around the police killing of Sean Bell in 2006 and protests and riots in Oakland, California, in response to the police killing of Oscar Grant in 2009. The online group ColorOfChange.org organized protests by tens of thousands of people around the Jena 6, Black high school students charged with attempted murder after a school fight. The fight occurred after they were threatened at school with a hangman's noose. This incident raised issues of both hate crimes and police bias (Oliver et al. 2019).

There was also a massive wave of protests by Latino immigrants in 2006 around proposals to make it a felony to aid undocumented immigrants (Bada et al. 2006; Bloemraad and Trost 2008; Carey et al. 2014; Zepeda-Millán 2017).

The Obama Years

But, then, the election in 2008 of Barack Obama, the first Black president in U.S. history, led many Americans of all races to celebrate that the United States had finally gotten past its long history of racism. Even people who did not support Obama politically, including his opponent John McCain, said his election was a positive turning point in U.S. racial history. But other Whites felt threatened by a Black president with a Muslim name.

Racial language was used to oppose President Obama's plan for universal health insurance. Opponents had blocked President Bill Clinton's attempts to pass universal health insurance in 1993, but did not use racial language to do it. But in 2009, the conservative commentator, Glenn Beck, called President Obama's plan for universal health insurance "stealth reparations" for Black people (Beck 2009) and said Obama had a "deep hatred of White people" (CBS News 2009). Maine Governor Paul LePage also told a group of Republicans in 2013 that President Obama "just hates White people" (Russell 2013), and racist images of President Obama circulated widely (Apel 2009; Joseph 2011). This racial language increased White opposition to policies from which White people would have benefitted and increased racial polarization and animosity.

The Tea Party faction within the Republican Party was originally organized by traditional economic conservatives and Republican Party leaders (Bakken 2014; Williamson et al. 2011). They protested in 2009–2010 against Obama's health care plan. White supremacist groups came to the Tea Party rallies to recruit new members (Zeskind 2012).

Whether it was motivated by simple partisanship or covert racism cannot be determined, but Republican Representatives and Senators adopted a strategy of complete noncooperation with President Obama. Although Democrats controlled both houses after the 2008 election, the House became Republican-controlled after the 2010 election as did the Senate after 2014. Contrary to custom, the Senate refused to ratify President Obama's judicial appointees. In a particularly egregious move, the Republican Senate refused to even hold hearings on President Obama's nominee for a Supreme Court

Justice in March 2016, seven months before the fall election and ten months before his term ended. The action's political cynicism was evident when Republican Senators rushed through the appointment of a Supreme Court justice a week before the fall 2020 election.

Members of political parties do what they can to increase voting by their supporters and, in some cases, to decrease voting by their opponents. If they have the power, they may also gerrymander, that is, draw voting district boundaries in ways that favor their Party. Black voters overwhelmingly support the Democratic Party, as do most voters from other racial minority groups and younger people. Older suburban and rural White people more strongly support the Republican Party. Efforts to make voting more difficult typically favor Republicans over Democrats. In recent years, Republican Party leaders have sought to reduce voting by Black voters and others who vote for Democrats. Some of these efforts are group-oriented, such as reducing the number of polling places in predominantly minority areas or gerrymandering district boundaries in ways that dilute the influence of minority voters. Others tie voting to things that are harder for lower-income or less-educated people to do, such as maintaining a stable address, obtaining a driver's license, or avoiding arrest. Because of the 15th Amendment, these efforts cannot be overtly discriminatory by race. The 1965 Voting Rights Act applied to states that had practiced voter racial discrimination and mandated certain voting protections. This law was renewed several times over the years. However, in 2013, the U.S. Supreme Court ruled that the Voting Rights Act was no longer needed and could not be enforced. The day after the Court decision, several states moved to restrict minority voting rights.

Black Lives Matter

The build-up toward Black Lives Matter began with the largely White Occupy Wall Street protests of 2011–2012. These had significant Black participation in New York, Oakland, and Atlanta (Taylor 2016), and there were spin-off Black "Occupy the Hood" movements in several cities (Grainger 2013; Morrow 2011; Paye 2012; Taylor 2016). There were protests about self-appointed neighborhood watchman George Zimmerman killing unarmed Trayvon Martin in 2012, and Zimmerman's 2013 not guilty verdict (Ransby 2015). There were large protests in New York in July 2014 about Eric Garner being choked to death by police officers. When Michael Brown was killed in Ferguson in August 2014, Black people used Twitter to rapidly spread the news that he was shot with his hands up, that his body lay in the street for four hours, and about the massive militarized police response to their initial peaceful protests (Bonilla and Rosa 2015; Randle 2016; Ray et al. 2017a, b; Thrasher 2015). Within a few days, activists had converged on Ferguson from all over the country. This was the take-off point for a wave of protests that spread around the country. The Twitter hashtag *#BlackLivesMatter* was created in 2013 on the day of the Zimmerman verdict by experienced community organizers Alicia Garza, Patrisse Cullors, and Opel Tometi. It became the name and slogan for the growing movement in November 2014, after the killing of Tamir Rice in Cleveland and the non-indictment of Darren Wilson, the officer who had killed Michael Brown (Anderson and Hitlin 2016).

The organization named Black Lives Matter was founded by the women who created the slogan. The protest movement using the slogan involved many different organizations, both long-standing community organizations and newly formed groups (Oliver 2020; Ransby 2015; Taylor 2016; Weddington 2018). Protest organizations combined national and local issues and linked police violence to broader issues of racism. The key organizations were Black-led, and some were restricted to Black-identified people in membership. Still, they received support from and worked closely in coalition with non-Black organizations and individuals. Many White people participated in "ally" organizations and protests in support of Black Lives Matter.

Forty organizations including Black Lives Matter met in Cleveland in 2015 to form the Movement for Black Lives (M4BL). They developed a broader agenda for social change https://m4bl.net/ that recognizes the shared struggles of all oppressed people and addresses issues of economic justice and political power in addition to policing. Their agenda was released originally in August 2016 and has continued to be developed since then.

Although many White people supported the Black Lives Movement,[6] others did not. The slogans "All Lives Matter" and "Blue Lives Matter" were deployed by opponents of the movement to detract attention from its core message. On July 7, 2016, as Black Lives activists were preparing the M4BL statement, a Black sniper shot 12 police officers, killing five of them, at a peaceful Black Lives Matter march in Dallas (Fernandez et al. 2016). The sniper supported the New Black Panther Party, a group that had been around since 1989 and was not part of M4BL. Several police chiefs and political

[6] See https://www.pediaa.com/what-is-the-difference-between-Black-lives-matter-and-movement-for-Black-lives/ Black Lives Matter is a specific organization, Movement for Black Lives (M4BL) is a coalition of organization which includes Black Lives Matter as one of the organizations. The slogan "Black Lives Matter" was used by many groups that were not members of the organization named Black Lives Matter. Many activists prefer the name Black Lives movement or Black Lives Movement to refer to the broader movement rather than one of the specific organizations in it.

leaders claimed that the Black Lives Movement was responsible for instigating anti-police violence. Many news stories described this shooting as the end of the Black Lives Movement.

The Trump Years

Donald Trump initially attracted national attention to himself by promoting "birtherism," the false belief that Obama was not a natural-born U.S. citizen.[7] Then, in 2016, he campaigned for the presidency by emphasizing anti-Muslim and anti-immigrant themes, opposed Black Lives movement, and police reforms. After his surprise victory in the electoral college despite losing the popular vote, White news media coverage and much public attention turned away from the Black Lives movement. White protests focused on the vast Women's marches against Donald Trump's election and inauguration. Once inaugurated, Donald Trump immediately launched initiatives to restrict entry into the United States of refugees and immigrants from many nations, actions which sparked a wide array of pro-immigrant protests. Many White people protested the escalating policies against legal immigrants and mistreatment of people seeking asylum in the United States, while other White people supported growing restrictions on immigration.

During the Trump years, the Movement for Black Lives and other organizations concerned with police violence and Black advancement kept organizing and forming coalitions with non-Black organizations. In August and September of 2016, NFL player Colin Kaepernick began kneeling during the National Anthem as a protest against police killings of unarmed Blacks. Team owners punished Kaepernick by not giving him a contract for the 2017 season. By 2017, other NFL players joined the "take a knee" protests, as did people in other settings. A crowd-sourced website of Black Lives protests[8] continued to list new protests every month between 2016 and 2020. Most of the original organizations are still operating, and new ones are forming all the time. They continued developing proposals and building coalitions

[7] Barack Obama was born in 1960 in Hawaii, a U.S. state since 1959, to a U.S. citizen mother and a Kenyan father. Birtherists promulgated the far-fetched belief that Obama's mother had actually flown to a place outside the United States to give birth.

[8] Alisa Robinson has collected news accounts of Black Lives Matter protests and organizations and policy statements and made them available through a creative commons license at https://elephrame.com/textbook/BLM/chart. She has also maintained an archive of organizations and policy statements List of BLM orgs https://elephrame.com/textbook/BLMOrganizations/chart List of BLM policy platforms https://elephrame.com/textbook/BLMPlatforms/chart summary of BLM published November 1, 2016 https://elephrame.com/textbook/What-is-the-Black-Lives-Matter-movement/article report about BLM published in 2017 https://elephrame.com/textbook/2017-Report-on-the-Black-Lives-Matter-movement/article.

and allies (Anspach 2018; Oliver 2020; Ransby 2018). M4BL sponsored the creation in 2017 of Rising Majority https://therisingmajority.com/, a coalition of groups led by Blacks and other people of color that links labor, youth, immigrant rights, climate change, feminist, anti-war, and economic justice groups. When George Floyd was murdered, they were ready to field protests and ready to talk about police reforms. They had built coalitions and located allies who were recently elected to local political offices. They had gained many White allies. Simultaneously, a number of White people increased their support for anti-racist movements. Others continued to fear Black protests and increased their support for police and White nationalists. Both sides mobilized around the 2020 Presidential election.

Meanwhile, a perfect storm was forming. White supremacists saw an opening in the Trump victory. They shocked many with their Unite the Right rally in Charlottesville, Virginia, in August 2017 featuring a march of young White men carrying Nazi symbols (Simi and Futrell 2020; Wellons and Jeffery 2017). In the years of the Trump administration, White supremacists and nationalists became more public in their actions. They began carrying assault weapons to protest COVID-control efforts in 2020 and menaced state legislatures warning them not to take actions they opposed. There were violent confrontations between them and anti-police protesters after the George Floyd killing. White supremacists even plotted to kidnap the governor of Michigan (Casiano 2020; Nicholas Bogel-Burroughs 2020) and harassed Black and other voters in the November election.

White supremacists had an ally in President Donald Trump. His verbal encouragement and refusal to condemn any of their acts gave them new life and legitimacy. They grew during the Trump years to an estimated 25,000 participants with about 150,000–175,000 supporters (Dobratz and Shanks Meile 1997; Simi and Futrell 2020). Then Trump provided them with a once in a lifetime opportunity. He was the first U.S. President to refuse to concede defeat in a national election, to falsely charge that the election was stolen from him (by fraudulent votes in key Black cities), and to call for Americans to come to Washington "to take back their government." Along with the unwillingness of Fox News and a number of key Republicans in Congress to challenge Trump's attempt to remain in office, the outcome was predictable. On January 6, 2021, all manner of right-wing White nationalist and racist extremists came to Washington and were encouraged in his speech before them to go to the Capitol. Some of them attacked the U.S. Congress with plans to kill, capture, and rout elected officials and prevent them from certifying the election of a new President, Joe Biden. This was the first act of violent sedition in the United States since the Civil War and only the second time the Capitol was violently damaged since the War of 1812. Anti-Black racism was a key motive.

Conclusions

There are two false narratives of U.S. racial history. First, the United States used to be racist when there was slavery or before the Civil Rights Movement, but there has been continuous progress since then. The other false narrative is that nothing has changed about U.S. racism in 150 years. A more accurate narrative is that history shows us that the structure of racial domination has always been changing in the struggle between racism and antiracism. Victories by one side lead to regrouping and a renewed effort by the other side.

Racism is about power and who makes the rules that affect everybody. It is a historical fact that White elites organized the United States to preserve a White majority at the expense of Black, Indigenous, and Asian people. Another fact is that there have always been challenges to this White domination. White people have disagreed about whether they benefit or are harmed by specific features of racism.

Indeed, White Americans have been affected by the intertwining of racism with other issues in politics. Racism affects White support for social policies that would benefit middle- and lower-income White people. The benefits include minimum wages, lower-cost public education, universal health insurance, income support for families with children, unemployment, disability, and old age supports. Intensive policing and mass incarceration, initially targeted on Black communities, have spread to lower-income White communities. The political coalitions that used White racial resentment to gain votes have allowed the political ascendancy of groups that many White people believe are dangerous to the well-being of most people in the United States, regardless of race. Using racial resentment is the same tactic deployed against Jews and other minorities by the German Nazi party to win elections and take power for Adolf Hitler. This kind of racist anti-minority politics is dangerous for everyone. In the wake of the COVID-19 pandemic, there are similar concerns. Anti-government and anti-tax policies have been promoted, in part, by racially coded language that has weakened essential government functions like public health to everyone's detriment, White people included. No individual has the power to make the past go away or to make racism disappear. But people can work together to push the system in the direction they want it to go and persuade others to join them.

Discussion Questions

Q1: In what instances have Whites opposed movements for greater racial equality or equity?

Q2: In what instances have Whites supported movements for greater racial equality or equity?

Q3: When have Whites benefited from their opposition to greater racial equality, and when have they benefited from their support of greater racial equality?

References

Aberbach, Joel D., and Jack L. Walker. 1970. "The Meanings of Black Power: A Comparison of White and Black Interpretations of a Political Slogan." *The American Political Science Review* 64 (2): 367–88. https://doi.org/10.2307/1953839.

Anderson, Monica, and Paul Hitlin. 2016. "Social Media Conversations About Race; How Social Media Users See, Share and Discuss Race and the Rise of Hashtags Like #Blacklivesmatter." http://www.pewinternet.org/2016/08/15/social-media-conversations-about-race/. *Pew Research Center*.

Anspach, Rachel. 2018. "The Movement for Black Lives Sharpens Focus in Trump Era." https://rewire.news/article/2018/09/13/the-movement-for-Black-lives-sharpens-focus-in-trump-era/. *Rewire.news*. (https://rewire.news/article/2018/09/13/the-movement-for-Black-lives-sharpens-focus-in-trump-era/).

Apel, Dora. 2009. "Just Joking? Chimps, Obama and Racial Stereotype." *Journal of Visual Culture* 8 (2): 134–42.

Bada, Xóchitl, Jonathan Fox, and Andrew Selee. 2006. "Preface." In *Invisible No More: Mexican Migrant Civic Participation in the United States,* Xóchitl Bada, Jonathan Fox and Andrew Selee, ed. Mexico Institute, Woodrow Wilson International Center for Scholars. Available from UC Santa Cruz Reprint Series, v–vi. https://escholarship.org/uc/item/7624m65m.

Bakken, P. Geoffrey. 2014. "'Let's Take Our Country Back!': The Tea Party Movement as a Moral Crusade." Ph.D., Sociology, University of Wisconsin—Madison, Madison, Wisconsin.

Balmer, Randall. 2014. "The Real Origins of the Religious Right." *Politico,* May 27.

Beck, Glenn. 2009. "Radio Clip." *Glennbeck.com.*

Bell, Joyce M. 2014. *The Black Power Movement and American Social Work*. New York: Columbia University Press.

Berbrier, Mitch. 1998a. "'Half the Battle' Cultural Resonance, Framing Processes, and Ethnic Affectations in Contemporary White Separatist Rhetoric*." *Social Problems* 45 (4): 431–50. https://doi.org/10.2307/3097206.

Berbrier, Mitch. 1998b. "White Supremacists and the (Pan-)Ethnic Imperative: On 'European-Americans' and 'White Student Unions'." *Sociological Inquiry* 68 (4): 498–516. https://doi.org/10.1111/j.1475-682X.1998.tb00482.x.

Berbrier, Mitch. 2000. "The Victim Ideology of White Supremacists and White Separatists in the United States." *Sociological Focus* 33 (2): 175–91. https://doi.org/10.1080/00380237.2000.10571164.

Berbrier, Mitch. 2002. "Making Minorities: Cultural Space, Stigma Transformation Frames, and the Categorical Status Claims of Deaf, Gay, and White Supremacist Activists in Late Twentieth Century America." *Sociological Forum* 17 (4): 553–91.

Berlet, Chip, and Stanislav Vysotsky. 2006. "Overview of U.S. White Supremacist Groups*." *Journal of Political and Military Sociology* 34 (1): 11–0_8.

Blee, Kathleen M. 1991. "Women in the 1920s' Ku Klux Klan Movement." *Feminist Studies* 17 (1): 57.

Blee, Kathleen M. 2005. "Racial Violence in the United States." *Ethnic and Racial Studies* 28 (4): 599–619. https://doi.org/10.1080/01419870500092423.

Blee, Kathleen M. 2017. "How the Study of White Supremacism Is Helped and Hindered by Social Movement Research." *Mobilization* 22 (1): 1–15. https://doi.org/10.17813/1086-671X-22-1-1.

Blee, Kathleen M., and Kimberly A. Creasap. 2010. "Conservative and Right-Wing Movements." *Annual Review of Sociology* 36 (1): 269–86. https://doi.org/10.1146/annurev.soc.012809.102602.

Bloemraad, Irene, and Christine Trost. 2008. "It's a Family Affair: Intergenerational Mobilization in the Spring 2006 Protests." *American Behavioral Scientist* 52 (4): 507–32.

Bloom, Joshua, and Waldo E. Martin. 2013. *Black Against Empire: The History and Politics of the Black Panther Party*. Berkeley: University of California Press.

Bonilla, Yarimar, and Jonathan Rosa. 2015. "#Ferguson: Digital Protest, Hashtag Ethnography, and the Racial Politics of Social Media in the United States." *American Ethnologist* 42 (1): 4–17. https://doi.org/10.1111/amet.12112.

Brewer, Rose M. 2003. "Black Radical Theory and Practice: Gender, Race, and Class." *Socialism and Democracy* 17 (1): 109–22. https://doi.org/10.1080/08854300308428344.

Browne, J. Zamgba. 1999. "Naacp Releases New Report on Profiling." *New York Amsterdam News* 90 (24): 3.

Burnham, Walter Dean. 1965. "The Changing Shape of the American Political Universe." *The American Political Science Review* 59 (1): 7–28. https://doi.org/10.2307/1976117.

Burris, V. A. L., Emery Smith, and A. N. N. Strahm. 2000. "White Supremacist Networks on the Internet." *Sociological Focus* 33 (2): 215–35.

Camp, Jordan T., and Christina Heatherton, eds. 2016. *Policing the Planet: Why the Policing Crisis Led to Black Lives Matter*. New York: Verso. https://search.library.wisc.edu/catalog/9912199693902121.

Carey, Tony E., Regina P. Branton, and Valerie Martinez-Ebers. 2014. "The Influence of Social Protests on Issue Salience Among Latinos." *Political Research Quarterly* 67 (3): 615–27.

Carroll, Leo, and Pamela Irving Jackson. 1982. "Minority Composition, Inequality and the Growth of Municipal Police Forces, 1960–71." *Sociological Focus* 15 (4): 327–46.

Casiano, Lous. 2020. "Disturbing New Details in Alleged Plot to Kidnap Michigan Gov. Gretchen Whitmer." *Fox News.*

CBS News. 2009. "Glenn Beck: Obama Is a Racist." *CBS News.*

Critical Resistance. 2018. "Critical Resistance Home Page." http://criticalresistance.org/. Retrieved 10/21/18.

Cunningham, David. 2003. "Understanding State Responses to Left-Versus Right–Wing Threats." *Social Science History* 27 (3): 327–70.

Cunningham, David. 2009. "Ambivalence and Control: State Action Against the Civil Rights-Era Ku Klux Klan." *Qualitative Sociology* 32 (4): 355–77. https://doi.org/10.1007/s11133-009-9137-1.

Cunningham, David. 2012. *Klansville, U.S.A.: The Rise and Fall of the Civil Rights-Era Ku Klux Klan.* Oxford University Press.

Davenport, Christian. 2010. *Media Bias, Perspective, and State Repression: The Black Panther Party.* Cambridge; New York: Cambridge University Press.

Davenport, Christian. 2012. "When Democracies Kill: Reflections from the Us, India, and Northern Ireland." *International Area Studies Review* 15 (1): 3–20. https://doi.org/10.1177/2233865912437149.

Davenport, Christian. 2015. *How Social Movements Die: Repression and Demobilization of the Republic of New Africa.* New York: Cambridge University Press.

Dello Buono, Richard Alan. 1992. "Criminalization of Black Radical/Nationalist Struggles: From the African Blood Brotherhood to Move." Paper presented at the American Sociological Association, 1992.

Dobratz, Betty, and Lisa Waldner. 2012. "Repertoires of Contention: White Separatist Views on the Use of Violence and Leaderless Resistance." *Mobilization: An International Quarterly* 17 (1): 49–66. https://doi.org/10.17813/maiq.17.1.328 2448743272632.

Dobratz, Betty A., and Lisa K. Waldner. 2006. "In Search of Understanding the White Power Movement: An Introduction*." *Journal of Political and Military Sociology* 34 (1): 1–0_8.

Dobratz, Betty A., and Stephanie Shanks Meile. 1997. '*White Power, White Pride!': The White Separatist Movement in the United States.* New York: Twayne Publishers; London: Prentice Hall International [1997] ©1997.

Fernandez, Manny, Richard Pérez-Peña, and Jonah Engel Bromwich. 2016. "Five Officers Killed as Payback, Chief Says." Pp. Section A; Column 0; National Desk; Pg. 1 in *The New York Times*.

Fletcher, Bill, Jr., and Jamala Rogers. 2014. "No One Said That It Would Be Easy." *Black Scholar* 44 (1): 86–112. https://doi.org/10.5816/Blackscholar.44.1.0086.

Francis, Megan Ming. 2014. *Civil Rights and the Making of the Modern American State*: Cambridge and New York: Cambridge University Press, 2014.

Giddings, Paula. 1984. *When and Where I Enter: The Impact of Black Women on Race and Sex in America*. New York: W. Morrow.

Giddings, Paula. 2008. *Ida: A Sword among Lions: Ida B. Wells and the Campaign Against Lynching*. New York, NY: Amistad.

Goldstein, Robert Justin. 2001. *Political Repression in Modern America from 1870 to 1976*. Urbana: University of Illinois Press.

Grainger, Garrett L. author. 2013. "The Struggle Within the Struggle: Neoliberalism and Group Schism Within a Multiracial Social Movement Organization."

Heideman, Paul. 2019. "The Most Important Black Radical You've Never Heard Of." *Jacobin*. https://www.jacobinmag.com/2019/06/Black-radicalism-hubert-harrison-web-dubois-malcolm-x.

Jackson, Pamela Irving, and Leo Carroll. 1981. "Race and the War on Crime: The Sociopolitical Determinants of Municipal Police Expenditures in 90 Non-Southern U.S. Cities." *American Sociological Review* 46 (3): 290–305.

Johnson, Cedric. 2007. *Revolutionaries to Race Leaders Black Power and the Making of African American Politics*. Minneapolis: University of Minnesota Press, c2007.

Jones, Martha S. 2020. *Vanguard: How Black Women Broke Barriers, Won the Vote, and Insisted on Equality for All*: Hachette UK.

Joseph, Ralina L. 2011. "Imagining Obama: Reading Overtly and Inferentially Racist Images of Our 44th President, 2007–2008." *Communication Studies* 62 (4): 389–405. https://doi.org/10.1080/10510974.2011.588074.

Kelley, Stanley, Richard E. Ayres, and William G. Bowen. 1967. "Registration and Voting: Putting First Things First." *The American Political Science Review* 61 (2): 359–79. https://doi.org/10.2307/1953251.

Kerner, Otto, John V. Lindsay, and F. R. Harris. 1968. *Report of the National Advisory Commission on Civil Disorders*. Washington, DC: US Government Printing Office.

Kwoba, Brian. 2017. "Hubert Harrison: Black Griot of the Harlem Renaissance." *Black Perspectives*.

Louis Harris and Associates Inc. 1968, "Harris 1968 National Malaise Survey, No. 1813. Hdl:1902.29/H-1813." Odum Institute [distributor (DDI)]. https://dataverse.unc.edu/dataset.xhtml?persistentId=hdl:1902.29/H-1813.

Mansbridge, Jane J. 1986. *Why We Lost the Era*. Chicago: University of Chicago Press.

Marx, Gary T. 1970. "Civil Disorder and the Agents of Social Control." *The Journal of Social Issues* 26 (1): 19–57.

Marx, Gary T. 1974. "Thoughts on a Neglected Category of Social Movement Participant: The Agent Provocateur and the Informant." *American Journal of Sociology* 80 (2): 402–42.

Marx, Gary T. 1980. "The New Police Undercover Work." *Urban Life* 8 (4): 399–446.

Marx, Gary T. 1981. "Ironies of Social Control: Authorities as Contributors to Deviance Through Escalation, Nonenforcement and Covert Facilitation." *Social Problems* 28 (3): 221–46.

Marx, Gary T. 1982. "Who Really Gets Stung? Some Issues Raised by the New Police Undercover Work." *Crime and Delinquency* 28 (2): 165–93.

Mayer, Jane. 2017. *Dark Money: The Hidden History of the Billionnaires Behind the Rise of the Radical Right*. New York: Anchor Books.

Migration Policy Institute. 2020, "U.S. Immigration Trends." https://www.migrationpolicy.org/programs/data-hub/us-immigration-trends.

Morrow, Christian. 2011. "'Occupy the Hood' Expanding." In *New Pittsburgh Courier*, 2-A1, A4. Pittsburgh, PA.

Nicholas Bogel-Burroughs. 2020. "What We Know About the Alleged Plot to Kidnap Michigan's Governor." *New York Times*. https://www.nytimes.com/2020/10/09/us/michigan-militia-whitmer.html.

Oliver, Pamela. 2008. "Repression and Crime Control: Why Social Movement Scholars Should Pay Attention to Mass Incarceration as a Form of Repression." *Mobilization: An International Quarterly* 13 (1): 1–24.

Oliver, Pamela. 2020. "Resisting Repression: The Black Lives Movement in Context." In *Racialized Protest and the State: Resistance and Repression in a Divided America*, edited by H. Johnston and P. Oliver, 63–88. Oxford, UK and New York, NY: Routledge.

Oliver, Pamela E., Alex Hanna, and Chaeyoon Lim. 2019. "Black Protest Events in the Us 1994–2010: Issues, Campaigns and Trends." Available from SocArXiv at https://osf.io/preprints/socarxiv/aq5xj.

Paul, M. Heideman. 2018. *Class Struggle and the Color Line: American Socialism and the Race Question 1900–1930*. Chicago, IL: Haymarket Books, 2018. ©2018.

Paye, Amity. 2012. "Occupy the Hood Makes Big Plans During Hood Week Atlanta." *New York Amsterdam News*, 4. New York, NY.

Perry, Jeffrey B. 2009. *Hubert Harrison the Voice of Harlem Radicalism, 1883–1918*: Columbia University Press.

Pierce, Paulette. 1988. "The Roots of the Rainbow Coalition." *The Black Scholar* 19 (2): 2–16.

Pierce, Paulette. 2003. "Neglected Legacy: The Black Power Movement and Jesse Jackson's Presidential Campaigns." *Humanity and Society* 27 (4): 481–97. https://doi.org/10.1177/016059760302700406.

Piven, Frances Fox. 1988. *Why Americans Don't Vote*. 1st ed. New York: Pantheon Books, [1988] ©1988.

Randle, Aaron. 2016. "Now You See Me: A Look at the World of Activist Johnetta Elzie." *Complex.com*, March 8.

Ransby, Barbara. 2015. "The Class Politics of Black Lives Matter." *Dissent*, Fall, 31–34.

Ransby, Barbara. 2018. *Making All Black Lives Matter: Reimagining Freedom in the Twenty-First Century*. Oakland, CA: University of California Press, [2018] ©2018.

Ray, Rashawn, Melissa Brown, Neil Fraistat, and Edward Summers. 2017a. "Ferguson and the Death of Michael Brown on Twitter: #Blacklivesmatter, #Tcot, and the Evolution of Collective Identities." *Ethnic and Racial Studies* 40 (11): 1797–813. https://doi.org/10.1080/01419870.2017.1335422.

Ray, Rashawn, Melissa Brown, and Wendy Laybourn. 2017b. "The Evolution of #Blacklivesmatter on Twitter: Social Movements, Big Data, and Race." *Ethnic and Racial Studies* 40 (11): 1795–96. https://doi.org/10.1080/01419870.2017.1335423.

Richie, Beth. 2012. *Arrested Justice Black Women, Violence, and America's Prison Nation*: New York: New York University Press, c2012.

Robnett, Belinda. 2020. "Defensive Adaptations: Naacp Responses to the U.S. Post Racial Project 1970–1990." In *Racialized Resistance and the State: Resistance and Repression in a Divided America*, edited by H. Johnston and P. Oliver, 26–62. Oxford, UK and New York, NY: Routledge.

Russell, Eric. 2013. "Republican Lawmakers: Lepage Said Obama 'Hates White People'." *Portland (Maine) Press Herald*. Portland, Maine. https://www.pressherald.com/2013/08/19/lepage-is-heard-to-say-obama-hates-White-people/.

Simi, Pete, and Robert Futrell. 2020. "Active Abeyance, Political Opportunity, and the 'New' White Supremacy." In *Racialized Protest and the State: Resistance and Repression in a Divided America*, edited by H. Johnston and P. Oliver, 112–39. Oxford UK and New York NY: Routledge.

Stillerman, Joel. 2020. "The Negro Labor Committee and the Dilemmas of Black Trade Unionism in New York City, 1925–1950."

Taylor, Keeanga-Yamahtta. 2016. *From #Blacklivesmatter to Black Liberation*. Chicago, IL: Haymarket Books.

Theoharis, Jeanne. 2018. *A More Beautiful and Terrible History: The Uses and Misuses of Civil Rights History*. Boston, MA: Beacon Press.

Theoharis, Jeanne, and Komozi Woodard, eds. 2005. *Groundwork: Local Black Freedom Movements in America*. New York: New York University (Table of contents http://www.loc.gov/catdir/toc/ecip0420/2004017053.html).

Theoharis, Jeanne F., and Komozi Woodard, eds. 2003. *Freedom North: Black Freedom Struggles Outside the South, 1940–1980*. New York: Palgrave Macmillan.

Thrasher, Steven W. 2015. "What Next for Black Lives Matter in Ferguson After City's Police Shooting?" *The Guardian (US Edition)*. https://www.theguardian.com/us-news/2015/mar/13/Black-lives-matter-ferguson-police-shooting.

Upton, Aisha A., and Joyce Bell. 2017. "Women's Activism in the Modern Movement for Black Liberation." In *The Oxford Handbook of U.S. Women's Social Movement Activism*, edited by H. J. McCammon, V. Taylor, J. Reger, and R. L. Einwohner: Oxford Handbooks Online.

Urban America and Urban Coalition. 1969. *One Year Later; an Assessment of the Nation's Response to the Crisis Described by the National Advisory Commission on Civil Disorders*. New York: Praeger.

Weddington, George. 2018. "Organizing Against Incarceration: Foucault's 'Carceral Continuum'." Paper presented at the Association of Black Sociologists, Philadelphia, PA.

Wellons, Mary Catherine, and Adam Jeffery. 2017. "Scenes from Charlottesville's Violent 'Unite the Right' Rally." *CNBC.com*.

White, Deborah Gray. 1999. *Too Heavy a Load: Black Women in Defense of Themselves, 1894–1994*. 1st ed. New York: W. W. Norton, 1999. ©1999.

Williamson, Vanessa, Theda Skocpol, and John Coggin. 2011. "The Tea Party and the Remaking of Republican Conservatism." *Perspectives on Politics* 9 (1): 25–43.

Zepeda-Millán, Chris. 2017. *Latino Mass Mobilization: Immigration, Racialization, and Activism*. Cambridge: Cambridge University Press.

Zeskind, Leonard. 2012. "A Nation Dispossessed: The Tea Party Movement and Race." *Critical Sociology* 38 (4): 495–509. https://doi.org/10.1177/0896920511431852.

13

Affirmative Action: Not the Harsh Impact on Whites That Some Assume

Fred L. Pincus

Abstract Fred Pincus explains what affirmative action is. He dispels the conservative truism that affirmative action is about reverse discrimination against White Americans. Affirmative action has benefited the overwhelming majority of people who have ever applied for work or applied to a college or university because it established a procedure for fairness and makes decisions less arbitrary. Intentional interventions into the racial practices and outcomes of institutions are the only effective way to correct historic and contemporary racial inequalities. New affirmative action-type measures will have to be taken to reduce racial and other inequalities.

Keywords Affirmative action · Racial discrimination · Employment · Higher education · Racial inequality · White Americans

Affirmative action is one of the most contentious and misunderstood policies intended to create a more egalitarian society. There is disagreement about what constitutes affirmative action, whether it is effective, and whether it is just or fair. Many Whites believe that they, as a group, are harmed by race-based affirmative action. Some of this disagreement can be addressed by

F. L. Pincus (✉)
Department of Sociology, University of Maryland, Baltimore County, Baltimore, MD, USA
e-mail: pincus@umbc.edu

D. W. Austin and B. P. Bowser (eds.), *Impacts of Racism on White Americans In the Age of Trump*, https://doi.org/10.1007/978-3-030-75232-3_13

examining quantitative data and legal regulations (Kiely and Marcuis 2018; Pincus 1993, 1999, 2003). But much of the controversy goes beyond facts and involves politics, philosophy, and emotion for which there are no correct and incorrect answers.

The affirmative action debate precedes President Donald Trump's time in office and will likely continue afterward. However, his conservative attacks on federal regulations and his not-so-subtle attacks on people of color and support of White supremacists do not bode well for affirmative action. In one of the major affirmative action cases to come before the federal courts during his term, the Trump administration took the side of those criticizing affirmative action in college admissions. This case will be discussed in a later section.

Much of the opposition to affirmative action denies the existence of systemic racism and White privilege. In her study of White Tea Party supporters in Louisiana before the 2016 election, Arlie Russell Hochschild (2016) uses the metaphor of White people patiently waiting in line to achieve the American Dream that is just over the hill. They have worked hard and followed the rules. They are not bothered when someone before them lets a friend, relative, or coworker cut in the line in front of them; this is seen as legitimate networking. However, suppose the government allows a person of color, who is less likely to have the same networks, to cut into the line in front of them. In that case, Whites would get angry and see this as illegitimate, unfair, and an example of reverse discrimination.

This way of thinking is partially due to the American myth that anyone who works hard can get ahead, regardless of barriers they may confront. This myth is reinforced by conservative think tanks and legal groups that have opposed affirmative action for decades. Systemic racism has no place in this way of thinking.

Since there is so much misinformation about affirmative action, I want to spend most of this chapter looking at the facts: What is affirmative action? What are attitudes toward affirmative action? Does affirmative action help people of color and women? Does affirmative action hurt Whites? I will end by addressing the broader political question: What should the future of affirmative action be?

One caveat before proceeding. Affirmative action alone will not solve the problem of racial inequality in employment or higher education. It should be an essential part of an anti-racism tool kit that can be used against structural racism and economic inequality. I will return to this point at the end of the chapter as well.

What Is Affirmative Action?

Many important policies are intended to be "color blind" and anti-discriminatory in that they mandate treating Whites and people of color the same. Not hiring someone because they are Black or not admitting someone to a college because they are Latinx would be considered racial discrimination. Civil rights laws require employers and colleges to treat Whites and people of color equally. However, if these nondiscrimination policies resulted in few Blacks hired or few Latinxs admitted to college, that would not necessarily be unfair or illegal. If employers found that fewer Blacks were qualified or if college admissions officers found that fewer Latinxs had the necessary grades and test scores, so be it. Meritocratic requirements (i.e., the highest-scoring people get the job or college seat) are required. In other words, treat people equally and let the chips fall where they may.

However desirable and necessary color-blind and anti-discriminatory laws are, they do not constitute affirmative action. *Affirmative action policies take race into account in a variety of ways. They go beyond meritocratic, color-blind, anti-discrimination laws and policies.* In the labor force, this often involves setting goals and timetables to increase people of color's employment. In higher education, it usually consists of using race as one of many factors in college admission. I will refer to this as the "race-plus" policy.

It should come as no surprise that even the terminology about affirmative action is contentious. For example, some writers talk about "soft" vs. "hard" affirmative action. Their version of "soft" affirmative action refers to what I have called "anti-discrimination laws" above. For example, Melvin Urofsky (2020) refers to soft affirmative action as "programs that would open the doors of opportunity – whether in industry, education, or government – to minorities that have hitherto been excluded. This would be achieved in part by federal law to ensure that national, state, and local governments as well as private employers did not discriminate against minorities" (2020, 57).

"Hard" affirmative action, on the other hand, would refer to those programs that go beyond color-blind policies in that they take race into account. Many of these hard/soft writers come to the same conclusion that Urofsky does: "I strongly favor what I have called the 'soft' version ... I am not enamored of the 'hard' version which is numbers-driven" (2020, xvii).

My translation of Urofsky and his colleagues is that they like anti-discrimination laws, but they do not like affirmative action. While everyone has a right to their own definitions, I will focus on those policies that go beyond anti-discrimination laws.

Many people mistakenly equate "goals" and the "race-plus policy" with "quotas." In the first page of Urofsky's recent book, he says "The use of quotas, even if the firms often used the euphemism of 'goals,' offended many people" (2020, ix). Yet, the term "goals" does not even appear in the index of his book, although "quotas" do. As we shall see, goals and quotas are *not* the same thing and the use of quotas in affirmative action is rare.

While the theme of this book is how racism impacts both White people and people of color, any discussion of affirmative action must also consider gender. White men and women might well be impacted in different ways. In terms of college admissions, as we shall see, Asians, in addition to Whites, are an advantaged group in that they are overrepresented in higher education. So, it is sometimes necessary to compare Whites and Asians, on the one hand, to non-Asian people of color, on the other.

In this chapter, I will focus on two types of affirmative action after discussing attitudes toward affirmative action. First, I will discuss the affirmative action policies in the labor market that are overseen by the Office of Federal Contract Compliance Programs (OFCCP) of the U.S. Department of Labor. Next, I will examine the admission policies for institutions of higher education.

Attitudes Toward Affirmative Action

Public opinion polls have surveyed Americans about affirmative action for many years although the results are often contradictory. First, the phrasing of the questions has a great impact on the results. On the one hand, respondents are supportive of affirmative action programs that help people of color. On the other hand, when asked if they support racial preferences as opposed to merit, respondents strongly support merit.

The data in Table 13.1 show that when asked a general question about support for affirmative action for both women and minorities, respondents respond increasingly positive between 2001 and 2018. Women are more supportive than men. Blacks and Latinxs are more supportive than Whites.

The same trend occurs when respondents are asked if they think affirmative action to increase the number of Black and minority students on college campuses is a good thing or bad thing. In 2017, 71% said it was a good thing, an 11% increase from 2003.

However, when asked how important race should be in college admissions decisions, almost three-quarters of respondents say it should not be considered at all (see Table 13.2). Even strong majorities of Blacks and Latinxs say race should not be considered at all. Whites and Republicans are especially

Table 13.1 Percent of adult population that support affirmative action for women and minorities

Target of affirmative action	Year	
	2001	2018
For women	53	65
Men's answers	49	61
Women's answers	57	69
For minorities	47	61
White's answers	44	57
Black's answers	69	72
Latinx's answers	64	66

Note 57% of Whites supported affirmative action for minorities in 2018 compared to 72% of Blacks (Norman 2019)

Table 13.2 Percentage saying race or ethnicity should be a ____ in college admissions decisions in 2019

Type of respondent	Importance of race or ethnicity			
	Major factor	Minor factor	Not a factor	Total
All adults	7	19	73	100
Race/ethnicity				
Whites	4	18	78	100
Blacks	18	20	62	100
Latinxs	11	22	65	100
Asians	13	29	58	100
Political affiliation				
Republican	4	12	85	100
Democratic	10	26	63	100

Note 78% of Whites thought race should not be a factor in college admissions compared with 62% of Blacks (Graf 2019)

opposed to using race. One particularly interesting finding is that Asians, who have grades and test scores that are even higher than Whites, are the *least* hostile to using race in the admissions process.

The strong White opposition to using race as a factor in college admission is consistent with the growth of feelings of White victimization. A 2017 survey shows that 55% of Whites believe that Whites are discriminated against. Eleven percent say that they, personally, have been discriminated against in college admissions and 19% say they have been discriminated against in applying for jobs (Gonyea 2017).

These ambivalent attitudes toward affirmative action, along with the belief in White victimization, should be viewed in the context of a conservative backlash against all civil rights policies since the 1960s. *U.S News and World Report, the Harvard Business Review and the Rutgers Law Review* all carried articles on the "dangers of reverse discrimination" as far back as the late 1960s. Conservative foundations like the Institute for Educational Affairs and the Earhart Foundation supported anti-affirmative action scholars in the 1980s (Pincus 2003). The Center for Individual Rights, which brought lawsuits against the University of Texas (see below) has received funding from a host of right-wing foundations including Bradly, Olin, Scaife, and Donors Trust and also has ties to Koch Brothers organizations (SourceWatch 2017, 2019).

Students for Fair Admissions (SFFA), who sued Harvard and the University of North Carolina continues the tradition (see below). Although it claims to be a pro-Asian student organization with a $10 membership fee, the *Harvard Crimson* found that SFFA received six-figure grants from the Searle Freedom Trust and Donor's Trust that totaled $750,000 in 2016 (Caldera 2019).

These and other organizations promote hostility toward affirmative action, misinformation about what affirmative action is, and the belief in White victimization. The next section begins to describe the reality of affirmative action.

Affirmative Action in Employment

The largest federal affirmative action program is based on Executive Order 11246 issued by President Lyndon B. Johnson in 1965. Guidelines to implement this program were first issued in 1968 and revised in 1971. Federal contractors and subcontractors, excluding those in construction, who have 50 or more employees and a $50,000+ federal contract are required to develop an affirmative action plan within 120 days of receiving a contract. Failure to develop and implement an affirmative action plan could result in a contractor losing the current contract and being declared ineligible to receive additional contracts. This is called being "debarred."

These guidelines apply only to a select group of federal contractors. Smaller federal contractors and most non-contractor corporations and businesses, large and small, are not required to have affirmative action plans. *Most employers in the United States are not subject to affirmative action regulations.*

Contractors meeting the 50/$50,000 criteria must first conduct a *utilization study* of their employees. Basically, they must count the number of women, people of color, veterans, and disabled people in each department and in each occupational category.[1]

The employer must also determine the percentage of people of color and female employees who are in the "availability pool"; i.e., those outside the company who qualified and are potentially available for the job. This is a complex issue and requires some explanation.

For most less skilled clerical, sales, blue-collar, and service jobs, availability is calculated as the percentage of people of color or women in the surrounding labor force. In more skilled jobs like carpenters, on the other hand, the availability would be the percentage of people of color or women workers employed in that job in the immediate geographical area.

For professional and managerial jobs, however, the availability may well be state-wide or even national. For nurses, for example, the Black availability might be defined as the percentage of Blacks getting nursing degrees in the entire state in the last five years. For college faculty, on the other hand, the availability of female mathematicians might be the percentage of Ph.D.'s in mathematics granted to women in the past five years in the entire country.

There are pages of regulations specifying how these figures are calculated. The important point here is that the availability pool is an estimate of the percentage of *qualified* people of color and female workers in a job category.

Employers must then compare their actual distribution of people of color or female employees in a specific job category in a specific department to the people of color and female distribution in the availability pool. If the actual employment is equal to or greater than the availability, the employer is "in compliance." If, on the other hand, the actual employment distribution is below the availability figure, the employer is "underutilized." The employer must follow this same procedure for each job category in each department.

Contractors and Affirmative Action

If a contractor is underutilizing people of color and women, a set of goals and timetables must be included in the affirmative action plan. An appropriate goal would be to hire enough qualified people of color and female employees to reach the percentage distribution stated in the availability pool.

The timetable must be based on the conditions facing that specific contractor. Employers with big turnovers might be able to reach the goal in a

---

[1] For the purposes of this chapter, I will not discuss veterans and the disabled. They deserve their own special discussion.

few months, while one with little turnover might take a few years. Contractors in expanding industries would have shorter timetables than those in stagnant or contracting industries.

Next, the contractor must specify procedures to achieve the goal. This "good faith effort" means trying to "cast a broad net" to encourage diversity among those who apply for the position. Employers should publicly advertise jobs rather than rely on informal networking. Advertisements should contain a statement like "Equal Opportunity Employer; Women and Minorities Encouraged to Apply." Some advertisements should be placed in publications targeted at qualified women and people of color. Employers should send letters to well-known women and people of color in the field asking for referrals, send letters to schools who train large numbers of qualified women and people of color, and make recruiting trips to conferences that might be attended by qualified women and people of color. In other words, employers must go out of their way to increase the hiring pool of potential women and people of color candidates.

The contractor is <u>not</u> required to submit the affirmative action plan to the OFCCP for approval. The plan must simply be kept on file in the contractor's office. What happens if the contractor fails to meet the goal specified in the plan? Probably nothing.

First, only the employer is likely to know that the goal was not met since the OFCCP does not review the hiring process of each employer for each year. In the unlikely event that the employer is ever investigated by the OFCCP, all the contractor would have to do is submit the plan and show that they followed the procedures to encourage women and people of color to apply for the position. If the contractor can demonstrate that the White male who applied for the administrative position was more qualified than any of the Black and female applicants, there is no problem. Affirmative action guidelines *require* meritocratic hiring. Mandatory or voluntary quotas are *illegal* under these guidelines.

Although many affirmative action critics believe that goals and quotas are the same thing, there are important differences. Most importantly, court-imposed quotas must be met. Goals require only a good faith effort. This will be discussed further below.

While these affirmative action regulations involve a certain amount of effort and cost on the part of federal contractors, *they do not force contractors to hire unqualified people, nor do they permit quotas*, apart from those that may have been mandated by a court. Some contractors, however, may pressure personnel officers to illegally hire unqualified underutilized minorities to avoid problems with OFCCP officials. It is difficult to determine how extensive this practice may be.

Construction contractors are also required to establish goals and timetables. However, they are not required to have full affirmative action plans on file because they do not have the same kind of stable labor force as a manufacturer might have. Many construction contractors hire different people from one job to the next.

The OFCCP does conduct "compliance reviews" of certain contractors who are suspected of not fully complying with guidelines. The number of annual reviews has been sharply declining—6232 in 1989, 4162 in 2000, 2345 in 2015, and only 1331 in 2019. In two-thirds of these reviews, the contractor agreed to change some aspect of the affirmative action plan to bring it into compliance (Pincus 1993; Program Office, n.d., 2002; Urofsky 2020). These are called "conciliation agreements."

Although 1331 may seem like many reviews, it is important to remember that there were over 133,700 prime contractors in 2017 and an even larger number of subcontractors that fell under the OFCCP guidelines. At the rate of 1331 reviews each year, it would take the OFCCP *more than 100 years* to review all contractors even once. Consequently, contractors do not really have to worry very much about being reviewed. The number of race/ethnicity/gender conciliation agreements and consent decrees dropped from 297 in 2015 to 184 in 2019.

If the compliance officer and the contractor cannot reach an agreement, there are several levels of appeal available to the contractor. Recalcitrant contractors can ultimately be debarred; i.e., they can lose their existing contracts and be declared ineligible to receive future contracts. However, this is extremely rare.

According to the OFCCP, only 43 contractors had been debarred in the first 37 years of the agency's history through 2002, about 1.2 debarments per year. These 43 companies account for a tiny fraction of the more than half-million companies that have been government contractors since 1972. Twenty-five of the companies were declared ineligible during the 1972–1980 period which covered the Nixon, Ford, and Carter administrations. Four were declared ineligible during the Reagan years (1981–1988), three during the Bush Sr. years (1989–1992), eight during the Clinton years (1993–2000) and one during the first two Bush Jr. years.

A Freedom of Information Act request to get more recent data found only five companies that had been debarred between 2009 and 2019, a rate of less than 0.5 per year. This is a dramatic decline from previous decades.

What does a government contractor have to do to be debarred? The official list provides this information on 40 of the 43 companies debarred between 1972 and 2002. Half of the companies flagrantly violated the OFCCP regulations by refusing to develop an affirmative action plan or refusing to submit required statistical information. The remaining half were debarred for refusing to make good faith efforts to implement goals and timetables or by not abiding by some other part of the conciliation agreement.

The reality is that the federal affirmative action regulations that are administered by the OFCCP do not put a great deal of pressure on federal contractors to increase their hiring of women and minority workers. If a contractor is willing to be even the least bit flexible, the chances are good that the OFCCP will sign off on their affirmative action plans. Even after being debarred, companies can be reinstated if they make the necessary changes. Sixty percent of the companies debarred between 1972 and 2002 were eventually reinstated. In fact, the median period of debarment for the 26 reinstated contractors was only 9.5 months!

I have gone into so much detail about the OFCCP guidelines for several reasons. First, they are the least known of all affirmative action policies. Second, affirmative action critics usually do not discuss them. Third, they are the least controversial since the final hiring decisions are supposed to be meritocratic and companies are only required to *try* to meet their goals. Employers are only required to show that they have made a good faith effort to cast a broad net to find qualified candidates for their positions. Fourth, they affect more employees than other affirmative action policies.

Hiring and Promotion Quotas

The most controversial of all affirmative action policies are "quotas"; i.e., policies that reserve certain positions for qualified people of color or female candidates. This specifies a hiring or promotion floor; i.e., at least that number of women or minorities must be hired or promoted. This is different from the historic use of antisemitic quotas as a ceiling; i.e., no more than a certain number of Jews may be hired.

Although the concept of quotas is quite controversial, one thing is clear: quotas and goals are not the same. First, especially in court-imposed quotas, the employer or school must hire a person of color or female for that position, under penalty of law. If no qualified person of color or woman is found, either the position must remain empty or the employer must seek special permission to hire or promote a White male. In the case of goals, on the other hand, the employer must merely make a good faith effort to hire or promote a qualified person of color or woman; if none is found, there are no legal consequences and a White or male may be hired or promoted.

Second, in a consent decree (i.e., a legally binding agreement between two parties) with quotas, a White male with superior work experience or credentials, could be passed over in favor of a qualified, but less experienced person of color or female applicant. When goals are being used, a more qualified White male must be hired or promoted over a less qualified person of color or female applicant since the final hiring must be meritocratic. Affirmative action critics who say that goals and quotas are the same are either misinformed or intellectually dishonest.

The 1972 Equal Employment Opportunities Act made it legal for courts to impose hiring and promotion quotas on employers that were found guilty of discrimination. Quotas could also be used as a remedy as part of consent decrees, or out-of-court settlements in discrimination lawsuits.

Before a quota can be imposed by a court, a group of people of color or female employees generally sues an employer for race and/or sex discrimination. One possible outcome of these lawsuits is that the government and plaintiffs enter a consent decree with the employer that contains a quota system of hiring and/or promotion. The quotas might require the hiring of one Black for every five Whites until the percentage of Black employees reaches the level of Blacks in the availability pool.

Even under quotas, employers are not forced to hire unqualified people. Generally, the employer has some criteria which make the prospective employee "qualified"; e.g., an educational credential, a minimum score on a test, a minimum level of experience, etc. Employees who don't meet these criteria cannot be considered for the position. Hence, all those who do meet the criteria are seen as able to carry out the duties of the position.

Next, the qualified Whites (or males) are ranked from "most qualified" to "least qualified" in terms of the criteria. The same is done for the qualified people of color (or women). If only 10 people can be hired or promoted and there is a 50–50 quota, the top 5 Whites and the top 5 people of color are chosen. Even though all the selected people of color (or women) are qualified, it is possible that some of them may be less qualified (i.e., their scores are lower) than some of the Whites (or men) that were not chosen. For example, a woman who scored 85 out of 100 might be promoted while a man who scored 86 would not. While such a decision has a great impact on that particular man, such small differences in scores would not warrant the conclusion that the man would have done a better job than the woman.

There is a widespread belief, especially among Whites, that quotas are very common across the country. However, it is getting more and more difficult for hiring and promotion quotas to meet the "strict scrutiny" test of constitutionality. A variety of court decisions have concluded that there must be a "compelling state interest" to justify a quota. This is usually interpreted as the state has the duty to combat intentional race or gender discrimination when no other policy is likely to work.

Second, the quota system must be "narrowly tailored" which is generally interpreted as not "unduly trammeling" on the rights of White males. The consent decree cannot require that 100% of new hires be people of color or women since this would make it impossible for White males. In fact, the quota proportions must have some connection to the availability pool. In addition, the quota system cannot be in effect indefinitely; e.g., it may be in effect until the percentage of minority/female employees reaches a percentage equivalent to the availability pool.

In fact, court-imposed quotas are few and far between. Reskin (1998), in fact, says that there were only 51 court-approved quotas in effect in the early 1980s.

Non-contractor businesses can, of course, voluntarily adopt the goals-and-timetables type of affirmative action. In certain limited cases, non-contractors can adopt quota plans if they meet certain conditions: (1) a company has used hiring or promotion practices that are not job-related and have an adverse impact on people of color and/or women; (2) a company has used prior discriminatory practices; or (3) the labor pool is limited due to prior discrimination (Fick 2014; Service 2010). Of course, the voluntary affirmative action plan must also meet the strict scrutiny criteria outlined above.

Does Affirmative Action Work?

Does affirmative action increase the number of people of color and women employees? This is a complex question since it is necessary to disentangle the impact of simple anti-discrimination policies from affirmative action policies and from other employment and policy trends.

Women, for example, have increased their representation in the labor force and in predominantly male jobs, and the male–female income gap has declined somewhat. However, is this due to affirmative action, to sex discrimination lawsuits, to the increasing necessity for two-income families, to changes in the labor force, or to a combination of these and other factors?

I reviewed six different large-scale empirical studies that used multivariate analysis to answer this question. They cover the period from the late 1970s through the early 2000s. Two compared contractors and non-contractors (Kurtulus 2015; Leonard 1989); one looked at what happened with those who had just become contractors compared with those whose contracts had ended (Miller 2017); one looked at the results of a survey of employers (Holzer and Neumark, 1998); and two looked at major sex and race discrimination verdicts and settlements that contained required affirmative action regarding managerial occupations (Hirsh and Cha 2017, 2018).

While it is difficult to compare the findings from these diverse studies, I have tried to summarize them in Table 13.3. The authors of each study are listed in column 1 and the time period covered is listed in column 2. In columns 3–8, a "–" indicates a negative impact of affirmative action on that race/gender group; a "+" indicates a positive impact; a "0" indicates no impact or a negligible impact. There are several "not available" designations since not all the studies showed data by race/ethnic group. Miller, for example, talks about White women and all Blacks.

Looking down each column, Table 13.3 shows that both Black males and females benefit from affirmative action in three of the studies and "all Blacks" benefit in a fourth study. This is consistent with expectations. In the Kurtulus study, data also show that Native American men and women also benefit from affirmative action, Hispanic men are hurt, and Hispanic women, Asian women, and Asian men are not impacted. Summarizing previous studies

Table 13.3 Impact of affirmative action by race and gender in six multivariate studies

Authors	Years covered	Males		Females		Groups	
		White	Black	White	Black	All women	All Blacks
Hirsh and Cha (2018)	1997–2007	na	–	–	–	na	na
Hirsh and Cha (2017)	1996–2008	na	+	+	+	na	na
Miller (2017)	1978–2004	na	na	0	na	na	+
Leonard (1989)	1974–1980	na	+	0	+	na	na
Kurtulus (2015)	1973–2003	na	+	–	+	na	na
Holzer and Neumark (1998)	1992–1994	–	–	+	–	na	na

Note "+" means positive impact of affirmative action; "–" meant negative impact; "0" means no impact. Sometimes data were only reported for "all females" and/or "all Blacks." Studies not proving data for a race/ethnic group are indicated by "na"

of police, Garner et al. (2020) shows that Blacks are helped by affirmative action.

The impact of affirmative action on White women is less clear in Table 13.3. Two of the studies show positive impacts, two of the studies show negative impacts, and another two show no impact. The study on police mentioned above shows that "all women" benefit but we have no way to know how minorities faired because they did not use race/ethnic groups.

The lack of a clear positive impact on White women is surprising since it seems to be "common knowledge" that White women are beneficiaries of affirmative action. Many cite an Associated Press article in *The New York Times* discussing an untitled U.S. Labor Department 1995 report saying that "since the 1960s, affirmative action had helped five million members of minorities and six million women move up in the workplace." After repeated inquires, I finally located the report in the House Subcommittee on Employer-Employee Relations in 1995. In fact, the report simply showed the increase in the number of women in managerial, professional, sales, and clerical jobs between 1972 and 1992. Part of this increase was probably due to affirmative action, but we do not know by how much.

The available data show that affirmative action plans have positive impacts on the hiring of Black men and women. However, the impact on White women and other people of color is unclear.

Affirmative Action in Higher Education

Occupying a space somewhere between the goals/timetables approach and the quota approach is the race-plus policy. Race-plus policies permit race/gender to be considered as one of many factors in the final decision as long as it is not the *major* factor. Although employers can use the race-plus policy on a voluntary basis to rectify severe employment segregation, the race-plus policy has been more extensively used in higher education.

The underrepresentation of Blacks, Hispanics, and native Americans in higher education has been a major national issue for four decades. Since the late 1970s, colleges and universities around the country have used the race-plus principle that was articulated in the 1978 Bakke decision to help diversify their campuses (*Regents of the University of California v Bakke*, 1978). UC-Davis had set aside 18 seats for people of color and Alan Bakke, a White applicant who was not admitted, claimed that less qualified minorities were admitted. The Supreme Court found in favor of Bakke and said that the quota system was unconstitutional.

There were several important principles that resulted from this decision. First, the court ruled that although race cannot be considered as the *only* factor, it can be considered along with other factors such as economic disadvantage, athletic and music skills, parents' alumni status, etc. The court specifically cited Harvard University's race-plus undergraduate admissions policy as being constitutional. Second, the race-plus policy was not justified on the basis of compensating for past discrimination. Instead, Justice Lewis Powell said that a diverse student body contributes to a better educational experience for students through a more robust interchange of ideas. As a result of the Bakke decision, this race-plus policy has been used in admissions to undergraduate, graduate, and professional schools around the country.

Several developments during the mid-1990s led to restrictions on the use of the race-plus principle. In 1995, the Regents of the University of California voted to ban the use of race in admission and hiring. The following year, California voters approved Proposition 209 which banned the use of race as a factor in admissions, hiring, and promotions in all public institutions throughout the state. The U.S. Supreme Court declined to rule on a legal challenge to Proposition 209.

Also, in 1996, the Federal Appeals Court for the Fifth Circuit issued the *Hopwood vs. Texas* (1996) decision which involved the University of Texas law school. The court held that the race-plus principle was unconstitutional. Since the U.S. Supreme Court refused to rule on the case, none of the educational institutions in the Fifth Circuit (Texas, Mississippi, and Louisiana) could use the race-plus principle while the rest of the country could.

In 2003, the U.S. Supreme Court issued two important rulings relating to the University of Michigan. First, it said that giving prospective undergraduate candidates a certain number of points for being a person of color was unconstitutional. On the other hand, using race as one of many factors in the law school's "holistic" admissions process was constitutional (*Grutter v Bollinger*). This was reaffirmed by the Supreme Court in 2013 in a case sometimes referred to as Hopwood II.

Eight states have banned affirmative action in college admissions—California (1996), Washington (1998), Florida (1999), Michigan (2006), Nebraska (2008), Arizona (2010), New Hampshire (2011), and Oklahoma (2012). These bans were enacted through ballot initiatives, legislative referenda, court decisions, or state laws. Texas also banned affirmative action in 1996 but it was overturned by the Supreme Court in 2013. As we discussed in a previous section, conservative organizations promoted these bans including Students for Fair Admissions, the Center for Individual Rights, and the American Civil Rights Institute.

Baker (2019) used multivariate analysis to determine why one state would enact a ban while others would not. First, states with low percentages of White enrollment at flagship institutions were likely to ban affirmative action. This scarcity, Baker argued, caused the predominantly White population to feel threatened by race-plus affirmative action. Also, having a nearby state ban affirmative action increased the likelihood of neighbors instituting a ban.

In 1978, many affirmative action *supporters* viewed the Bakke decision's race-plus policy as a *defeat* for affirmative action because the court limited the use of voluntary quotas. There were numerous large demonstrations across the country criticizing the court. It is ironic that decades later, even the race-plus principle is under legal attack and affirmative action supporters are mounting a vigorous campaign to defend the Bakke decision that they once criticized as too narrow.

The Future of Affirmative Action

This brief review demonstrates how affirmative action policies in employment and higher education are supposed to work. Although I am a strong supporter of affirmative action, it is only one of many tools that can be used against racism in America. Larger issues of economic, structural, and cultural racism are beyond the scope of even the most effective affirmative action policies.

On the positive side, affirmative action makes it more difficult for individual college and employment recruiters to discriminate against anyone, White applicants included. It is harder for schools, employers, and government institutions to have discriminatory recruitment criteria at the institutional level. This, alone, is ample justification for keeping vibrant affirmative action policies.

On the other hand, the history of racism in the United States has created other barriers that go beyond affirmative action. For example, race and especially family income are both well-known predictors of educational attainment at all levels for a variety of reasons including the legacy of poor housing, poor schools, poor healthcare, low-paying jobs, and the high cost of college (Gilbert 2011). During the COVID-19 crisis, families got together with friends and neighbors to help their children navigate distance learning. Given the reality of race and class housing segregation and friendship networks, the more privileged White and upper-income students ended up getting the most support (Goodnough 2020). Larger structural changes are needed at all levels of society. Bernie Sanders' 2019 call for free college tuition would be a step in the right direction.

In my own city of Baltimore, for example, the prestigious Johns Hopkins University offers scholarships to any "high achieving" city resident that graduated from a city public school and meets the entrance requirements. The Baltimore Scholars Program offers a free ride for students from families with incomes of $80,000 or less and a partial scholarship for family incomes up to $150,000. In this predominantly Black city, only a handful of residents were able to qualify. Others could not overcome the race and class barriers that permeate Baltimore and other cities.

Getting back to affirmative action, per se, the almost fifty-year-old employment policies administered by the OFCCP have been institutionalized to some extent, even though implementation is weaker now than in the past. There is evidence from the studies I reviewed that many federal contractors implement these policies even when they are no longer required to do so because they are good business practices.

The Trump administration, to no one's surprise, aimed its deregulation and budget-cutting guns at the OFCCP. First it proposed to eliminate the OFCCP altogether by merging it into the Equal Employment Opportunity Commission although congress successfully resisted (Bertram 2018; Wilcher 2017). Next, the administration tried to slash the OFCCP budget in F2018 and 2019 but Congress also resisted that ploy. The number of OFCCP employees, however, declined to 496 in FY2020, fifty less than in FY2018.

Even more successful were the new regulations issued to weaken the OFCCP. One legal group that represents employers put it this way: "The good news for contractors is that the OFCCP's actions are almost all pro-business" (Kiely and Marcuis 2018). Only single-issue audits of contractors can be conducted meaning that if a contractor is suspected of discriminating against women, the disabled, and veterans, three separate audits must be conducted rather than one. The OFCCP was reminded to respect religious exceptions to discrimination. If a compliance evaluation ends with an Early Resolution Agreement, another compliance evaluation cannot be conducted for five years. These and other regulations make it more difficult for the OFCCP to do its job (Younies 2018). The goal may be to slowly chip away at employment-based affirmative action.

Affirmative action in higher education also faces an uncertain future. The lawsuit against Harvard University (*Students for Fair Admissions v Harvard*) alleging discrimination against Asians is critical. The plaintiffs allege that Harvard misuses its holistic review of prospective students by rating Asians low on personality characteristics, thus hurting their chances to be selected. Even though Asians are overrepresented at Harvard, the suit alleges that there is an informal quota to keep Asian enrollment from going any higher.

Consistent with its anti-regulation and anti-minorities political orientation, the Trump administration, of course, opposed Harvard's race-plus plan.

The lawsuit was brought by Students for Fair Admissions, a conservative organization that has also sued the University of North Carolina and other schools alleging discrimination against Whites and Asians. Although a federal district judge ruled in favor of Harvard in the fall of 2019, the case is widely expected to end up before the U.S. Supreme Court that currently has a conservative, anti-affirmative action majority.

Another challenge to affirmative action comes from critics who want to substitute color-blind policies to replace race-based affirmative action. Richard Kahlenberg (2014) and others have argued for years that since Blacks are overrepresented among low-income people, affirmative action that favors lower income students would act as a proxy for race; i.e., Black students would benefit from class-based affirmative action. In addition, class-based affirmative action would help to economically diversify college campuses which is also a major problem.

The empirical evidence comparing the two types of affirmative action is mixed. For example, two recent simulation studies argue that racial representation is substantially lower in class-based affirmative action than in race-based affirmative action (Reardon 2015; Xiang and Rubin 2015). On the other hand, other studies show that under certain circumstances, class-based plans can yield enrollment for people of color that is comparable to race-based plans (Gaertner and Hart 2015; Kahlenberg 2014).

Rather than taking an either/or approach, we should argue that both be done at the same time. After all, both race and class are predictive of academic success. The Reardon et. al. simulation study concludes that this combined race/class approach provides the biggest opportunity for students of color. Under no affirmative action plan, Black and Latinx enrollment would be a combined 5% of students. Compare this with 10% in a strong class-based plan, 26% in a strong race-based plan, and 38% in a strong combined plan. Unfortunately, few participants in this debate argue to add class-based affirmative action to the race-plus policies.

Another set of allegedly color-blind plans is what is usually referred to as the "top 10 percent plans." Texas, for example, accepts the top 10% of high school graduates to the University of Texas and Florida accepts the top 20%. Texas allows admission to the campus of the student's choice but Florida only guarantees admission to public university system. This top-percent plan takes advantage of the fact that most K-12 systems are highly segregated, so this is supposed to ensure a more adequate enrollment of Black and Latinx students without giving preference to any single race. In fact, the more segregated the

K-12 schools are, the larger the number of Black and Latinx students would be admitted. Whether or not the top-percent plans should be labeled "color blind" depends upon one's perspective.

Comparing these top 10% plans to more traditional race-based affirmative action policies is complex. In a review of the literature, Flores and Horn (Flores 2015) argue that the data for all three states show that Black and Latinx students are less likely to be enrolled in the *flagship* campuses than under race-based plans. The Black and Latinx enrollment gets redistributed to the less prestigious campuses. In Texas, the data are more favorable to the top 10% enrollment plan but Flores and Horn say that this is due to the rapid growth of the Latinx population, not to the plan itself (also see Kahlenberg 2014).

Klasik (2015) conducted a simulation of what a top 10% plan would look like in the University of Maryland system. Compared with a race-based admissions system, Klasik found that a top 10% plan would increase White and Latinx enrollment but decrease Black enrollment in Maryland.

Although these class-based affirmative action and top-percent plans are more likely to withstand legal scrutiny than the race-plus plans, it is likely that they will be less effective in terms of achieving racial equity in higher education. If the court strikes down race-plus plans in education, the decision might also negatively impact affirmative action in employment.

In terms of how affirmative action affects Whites, the empirical evidence is limited. Although half of Whites *believe* that anti-White discrimination is a serious problem, less than one-fifth believe that they, personally, have been discriminated against. Affirmative action is often blamed for much of this so-called reverse discrimination.

The empirical studies on hiring that were discussed in Table 13.3 showed inconsistent effects of affirmative action on White women and most did not measure the effects on White men. In a study going back two decades, Pincus (2003) found that the number of EEOC complaints and lawsuits filed by Whites alleging race-based employment discrimination was relatively small.

On the positive side, affirmative action policies weaken the impact of the "old-boy-network" to recruit employees and college students. By making recruitment policies public and subject to numerical review, affirmative action helps working-class Whites who tend to be less well connected than wealthy Whites who benefitted from the old-boys-network. This is often called "broadening the eligibility pool." Deemphasizing legacy admissions policies that primarily benefit the children of wealthy alumni would also benefit working-class Whites.

Seats at prestigious universities are scarce with rejection rates as high as 95%. When rejected, Whites can blame affirmative action without really knowing how admissions committees viewed their strengths and weaknesses. At many private elite universities, more legacy students, predominantly White, are admitted than affirmative action students but few people complain about that. Getting your kids into the same elite college that you attended is often seen as a justified privilege while affirmative action is not. The 2019 college admissions scandal added a new dimension to the role of privilege as 50 wealthy White families were accused of bribing officials at a variety of selective public and private universities to get their children admitted (Hearon 2020).

When you are in a crowded parking lot and see an empty space that turns out to be for the handicapped, you might think "If this weren't for handicapped, I could have gotten the space." This is doubtful, since someone else would have probably taken it. The problem is too many cars and not enough spaces. This is also the case for college admissions at elite universities.

Affirmative action increases competition for jobs and college seats which reduces White chances somewhat. In previous times, however, Blacks and Latinxs were not permitted to compete at all. Competition is part of the American Dream—except when White privilege is threatened.

Discussion Questions

Q1: Many Whites believe that affirmative action amounts to "reverse racism" or discrimination against White Americans. Does the data presented here support or reject that belief?

Q2: In what ways does affirmative action benefit White Americans?

Q3: In what ways does affirmative action reduce institutional racism?

References

Baker, D. J. 2019. "Pathways to Racial Equity in Higher Education: Modeling the Antecedents of State Affirmative Action Bans." *American Educational Research Journal* 56 (5): 1861–95.

Bertram, C. G. 2018. *Trump Administration Proposes Cuts to OFCCP Budget*, February 14. Retrieved from governmentcontractorcomplianceupdate.com.

Caldera, C. G. 2019. "Public Filings Reveal SFFA Mostly Funded by Conservative Trusts Searle Freedom Trust and Donors Trust." *Harvard Crimson*, February 7.

Fick, B. J. 2014. "The Case for Maintaining and Encouraging the Use of Voluntary Affirmative Action in Private Sector Employment." *Notre Dame Journal of Law, Ethics and Public Policy* 11: 15–70.

Flores, S. M. 2015. *Texas Top Ten Percent Plan: How It Works, What Are Its Limits, and Recommendations to Consider.* The Civil Rights Project.

Gaertner, M. N., and M. Hart. 2015. From Access to Success: Affirmative Action Outcomes in a Class-Based System. *University of Colorado Law Review*. https://scholar.law.colorado.edu/cgi/viewcontent.cgi?referer=https://www.google.com/&httpsredir=1&article=1047&context=articles.

Garner, M., A. Harvey, and H. Johnson. 2020. "Estimating Effects of Affirmative Action in Policing: A Replication and Extension." *International Review of Law and Economics* 62: 105881.

Gilbert, D. 2011. *The American Class Structure in an Age of Growing Inequality, Eighth Edition.* Thousand Oaks: Sage.

Gonyea, D. 2017. Majority of White Americans Say They Believe Whites Face Discrimination. www.npr.org/2017/10/24/559604836/majority-of-white-americans-think-theyre-discriminated-against/.

Goodnough, A. 2020. "They Can't Afford a Learning Pod. Now What?" *New York Times*, August 15, A1, A7.

Graf, N. 2019. *Most Americans Say Colleges Should Not Consider Race or Ethnicity in College Admissions*, February 25. Retrieved from PEW Research Center: www.pewresearch.org/fact-tank/2919/02/25.

Hearon, S. 2020. *The Biggest Names Associated With the College Admissions Scandal: Where Are They Now?* May 21. Retrieved from usmagazine: www.usmagazine.com/celebrity-news-pictures/college-admissions-scandal.

Hirsh, E., and Y. Cha. 2017. "Mandating Change: The Impact of Court-Ordered Policy Changes on Managerial Diversity." *ILR Review* 70 (1): 42–72.

Hirsh, E., and Y. Cha. 2018. "For Law and Markets: Employment Discrimination Lawsuits, Market Performance, and Managerial Diversity." *American Journal of Sociology* 123 (4): 1117–60.

Hochschild, A. R. 2016. *Strangers in Their Own Land: Anger and Mourning on the American Right.* New York: New Press.

Holzer, H. J., and D. Neumark. 1998. *What Does Affirmative Action Do?* Cambridge: National Bureau of Economic Research, Working Paper 6605.

Kahlenberg, R. D. 2014. *The Future of Affirmative Action: New Paths to Higher Education Diversity after Fisher v University of Texas.* New York: The Century Foundation Press.

Kiely, C. E., and Am J. Marcuis. 2018. *The OFCCP's Been Busy—9 New Directives This Year, Largely Pro-Business*, September 20. Retrieved from huntonlaborblog.com.

Klasik, D. 2015. *Can a Percent Plan Be a Successful Race-Neutral Alternative to Race-Conscious Affirmative Action in Maryland?* College Park: Maryland Equity Project.

Kurtulus, F. A. 2015. "The Impact of Affirmative Action on the Employment of Minorities and Women: A Longitudinal Analysis Using Three Decades of EEO-1 Filings." *Journal of Policy Analysis and Management* 35 (1): 34–66.

Leonard, J. S. 1989. "Women and Affirmative Action." *Journal of Economic Perspectives* 3 (1): 61–75.

Miller, C. 2017. "The Persistent Effect of Temporary Affirmative Action." *American Economic Journal: Applied Economics* 9 (3): 152–90.

Norman, J. 2019. *Americans' Support for Affirmative Action Programs Rises*, February 27. Retrieved from Gallup: news.gallup.com/poll/247046.

Pincus, F. L. 1993. "Enforcing Federal Affirmative Action Guidelines: Compliance Reviews and Debarment." *Journal of Intergroup Relations* 20: 3–11.

Pincus, F. L. 1999. "Reverse Discrimination: Fact and Fiction." *Perspectives: The Newsletter of Prejudice, Ethnoviolence and Social Policy* 7 (April/May): 1–6.

Pincus, F. L. 2003. *Reverse Discrimination: Dismantling the Myth*. Boulder: Lynne Rienner Publishers.

Program Office. 2002. *Facts on Executive Order 11246 Affirmative Action*. Retrieved from Office of Federal Contract Compliance Programs: www.dol.gov/esa/regs/compliance/ofccp/aa.htm.

Program Office. n.d. *Quick Facts*. Retrieved from Office of Federal Contract Compliance Programs: www.dol.gov/esa/media/reports/ofccp/ofqfacts.htm.

Reardon, S. F. 2015. *Can Socioeconomic Status Subsitute for Race in Affirmative Action College Admissions? Evidence From a Simulation Model*. Princeton: Educational Testing Service.

Reskin, B. 1998. *The Realities of Affirmative Action in Employment*. Washington, DC: American Sociological Association.

Service, C. R. 2010. *Affirmative Action in Employment: A Legal Overview*. Washington, DC: Congressional Research Service RL30470.

SourceWatch. 2017. *Center for Equal Opportunity*. Retrieved from SourceWatch: https://www.sourcewatch.org/index.php/Center_for_Individual_Rights.

SourceWatch. 2019. *Center for Individual Rights*. Retrieved from SourceWatch: https://www.sourcewatch.org/index.php/Center_for_Individual_Rights.

Urofsky, M. I. 2020. *The Affirmative Action Puzzle: A Living History from Reconstruction to Today*. New York: Pantheon Books.

Wilcher, S. 2017. *Trump Administration Proposes Merger of EEOC Under Guse of Efficiency*, August 15. Retrieved from insightsintodiversity.com.

Xiang, A., and D. B. Rubin. 2015. Assessing the Potential Impact of a Nationwide Class-Based Affirmative Action System. *Statistical Science* 30 (3): 297–327.

Younies, A. 2018. *OFCCP's Trump-era Audit Directives Combine Elements from Former Administrations*, December 9. Retrieved from hrunlimitedinc.com.

14

Summary: Racism's Impact on White Americans in the Age of Trump

Benjamin P. Bowser and Duke W. Austin

Abstract Each of the preceding chapters provided insights about White racism and its effects on Whites. The advantage of studying racism through an anthology of contributors is that the range of insights and expertise is wide enough to provide a big picture, a challenge for single authors to do. The disadvantage is that the book may seem fragmented. In this summary chapter, we review the contribution of each. We highlight contributors' insights and draw on the relevant findings of this book's two prior editions. Forty years of reflection and research are brought to bear. We summarize the impacts of racism on White Americans at this critical point in our nation's history—in the age of Trump.

Keywords White racism · Economy · Housing · Health · Government · Foreign affairs · Women and gender · Social psychology · Media · Sports · Education · Social movements · Affirmative action

B. P. Bowser · D. W. Austin (✉)
California State University East Bay, Hayward, CA, USA
e-mail: duke.austin@csueastbay.edu

B. P. Bowser
e-mail: benjamin.bowser@csueastbay.edu

D. W. Austin and B. P. Bowser (eds.), *Impacts of Racism on White Americans In the Age of Trump*, https://doi.org/10.1007/978-3-030-75232-3_14

Part I: Culture and Institutions

In the first edition of *Impacts,* Louis Kushnick (1981) asserted that cultural racism demands replication of White institutional dominance in employment, housing, schooling, government, criminal justice, the military, and even religion. However, the ultimate purpose for maintaining institutional control over people of color is to exploit White people. The rationale for using racism to control White people was articulated famously by former President Lyndon Johnson:

> If you can convince the lowest White man he's better than the best colored man, he won't notice you're picking his pocket. Hell, give him somebody to look down on, and he'll empty his pockets for you. (Knowles 1999, 408)

Maintaining "White" privilege over people of color distracts Whites Americans from appreciating that there are fellow Whites with vested interests in exploiting them in every way possible. Using race is necessary to distract Whites from using democratic government to restrain those who control them. Making Whiteness the basis of racial unity among people of European descent demands a scapegoat. In the long run, capitalism could not have advanced in the way it has without institutional racism. Ordinary White people would see who is "picking their pockets" and stop them.

Chapter 2: Economy

In the first edition of *Impacts*, the economist Michael Reich showed that reductions in racial and economic inequality are correlated with higher earnings for Whites and Blacks. In addition, reductions in inequality are also correlated with higher rates of unionization and falling industrial profits. He reaffirms these correlations in this edition. However, where racial inequality increases, low- and middle-income Whites lost financially, as during the Great Recession. Simultaneously, the upper one percent of White earnings increase and exceed the losses to the White majority. In the second edition of *Impacts*, Patrick Mason extended Michael Reich's analyses. His analyses showed that even without employer interference, conflict among workers lowered general wages and increased the number of people in the reserve labor force—those willing to work for less than those currently employed. With workplace racial hierarchy, more privileged White workers (though not all Whites) got slightly higher incomes, and less privileged Black workers earn lower incomes. Still, overall, all workers' wages and benefits are reduced while profits are increased.

Furthermore, the greater the need to constrain worker wages, address racial conflict, and increase production efficiency, the more managers and supervisors are needed. Middle managers, who form the American middle-class's backbone, control the labor force and minimize loss of overall performance and productivity. However, the salaries of middle-level employees reduce profits.

The question of who benefits from racial hierarchy upsets the trends described by Reich and Mason. Domestic elites internationalized U.S. capital and found foreign labor to work for much less, longer hours, and more efficiently. The initial outcome of moving work overseas became apparent in the 1960s and 1970s with deindustrialization, urban abandonment, and the resulting high levels of permanent Black unemployment. It took three decades for the same domestic disinvestment to catch up with White suburban and rural labor. Also, there is now less need for White middle managers to supervise a declining U.S. labor force. In 1996, Mason argued that globalization of U.S. capital and work was behind plant closings, deindustrialization, the de-skilling of American labor, decertification of unions, take-backs by management from unionized workers, and cuts in health and other "safety net" benefits. What first happened to Blacks is now happening to Whites; many Whites join Blacks in seeming to have no useful place as workers in neither the domestic nor international economy.

In all the editions of *Impacts*, we have learned that there are strong present-day economic incentives for Whites in the upper one percent of incomes to favor racial hierarchy. But there are also strong economic reasons for Whites in the lower and middle classes to do away with racism in the economy and in politics. Where institutional racism exists, White people are also harmed. And that harm to Whites runs through every major institution in American Life.

Chapter 3: Housing and Community

It was not until this third edition of *Impacts* that Jacob Rugh, in Chapter 3, demonstrated the earlier work of Kushnick, Reich, and Mason. Their historical observations have a present-day reality with social-political implications. According to Rugh, the 240 "low segregation-high education" counties in the United States represent the nation's most racially integrated and well-educated parts. On average, they are majority White, and closely reflect the nation's racial composition. These counties rejected Trump in 2016 and more so in 2020. Other counties that are quickly growing and becoming better educated and more racially integrated also rejected Trump (see the list of both sets of counties in Chapter 3). It is no coincidence that the number and sizes of the peaceful Charles Floyd and Black Lives Matter demonstrations occurred in the same counties.

Rugh suggests that another trend is emerging as well. There is new evidence that home values appreciate faster in stably integrated and diverse neighborhoods than in overwhelmingly White communities. Integrated spaces are now valued more highly than all-White spaces. To be socially mobile and a part of the economically fast-growing global world requires a racial and social exposure that is non-existent in White communities where residents are less educated and socially isolated. The 2016 and 2020 elections identified Trump supporters' geographic centers and rising White nationalism in less educated, highly segregated counties that are disproportionately White. Suggested here is a housing and community social ecology closely related to White decline and where Whites can be easily targeted with racist propagation. In contrast, those better able to resist racism cluster in rapidly growing, high education, racially integrated communities. Rugh identifies other costs of residential racial segregation for White Americans:

1. In the economy, *lower intergenerational social mobility* is significantly lower in more racially segregated metropolitan areas. Intergenerational social mobility is also more downward in White zip codes and for White people raised in White segregated areas (Chetty et al. 2014).
2. In jobs, *lagging economic growth* is slower in more segregated metropolitan areas, and the negative effect has increased over time alongside the shift from manufacturing to services and technology (Li et al. 2013). More racially integrated metro areas enjoy faster growth due to greater innovation fostered by more diverse entrepreneur networks and higher returns to venture capital (Samila and Sorenson 2017).
3. In Public Health, *Anti-vaccination beliefs* cluster spatially in lower educated, White segregated communities. Their racial homogeneity leads to a false sense that a highly contagious disease does not affect them and that mask-wearing is open to debate (Estep and Greenberg 2020). Politicized and racialized disbelief in science is concentrated disproportionately in White places.
4. In criminal justice, *fatalities involving police* kill 2.5 times more Black men than White men. This reality drives Black Lives Matters' calls for police reforms. However, fatality rates are elevated for White men in areas with a legacy of racial segregation and racial violence as in the South and the Second Amendment (gun) rural States. Police and gun reforms might

reduce the disproportion in the proportion of Black men killed, but it would benefit numerically more White men.

In this century, there is no single wealthy and steady growth nation without a strong government. It is impossible to effectively coordinate, manage, and regulate the affairs of large, diverse populations and economies without government on the same scale. Our time's central question is not government size, but how efficient and effective governments are and to what extent do they govern fairly, justly, equitably, and democratically? Yet in the United States, there has been a powerful conservative movement to reduce government size and limit its statecraft. Government size was debated among the United States founding fathers. The Southern slave plantation elite fought and lost a civil war for a small government that would not interfere in their affairs. The very same goal of a small and weak government has been the basis of a very effective conservative and libertarian resurgence since the 1981 Reagan Administration.

Why has the effort to reduce government been so successful in the United States despite small government being an anachronism? The answer to this question is racism. The fact is: there is a very close connection between the propagation of racism and successful advances in the conservative dream of a minimal federal government. The consequences of weakening the government have been negative for most Americans of all colors. However, here, we focus specifically on the results for White Americans. What happens when racism contributes to the crippling government?

Chapter 4: Health

In Chapter 4, Jonathan M. Metzl directly connects racism with the recent phenomenon of rising White death rates, declining health indices, and increasing poverty among Whites. The conventional explanation is the current heroin crisis is driving these declines among suburban and rural Whites and is caused by abusive prescription pain killers. Overprescription is only part of the story. The other part is that large segments of the White poor and working-class—Rugh's lower educated, segregated, and isolated Whites—have been led to oppose any government program that would benefit them. Through Fox News and other conservative commentaries, these programs are portrayed as benefiting Blacks and immigrants, not them. Racist commentary has generated White racial resentment to oppose government programs. So, Metzl interviewed Whites who would rather die from preexisting and preventable health conditions than get help and treatment from the government. Metzl points out that this racism is cultural and personal, and its impact is structural. For example, deaths due to racial resentment cluster in States that refuse to participate in the Affordable Care Act (ACA). The same racial fear historically and today has led to White gun ownership extremes, which have produced an ecology of rising White suicides and gun accidents. In contrast, States with the ACA and even modest gun control laws have significantly lower preventable White deaths and suicides.

Chapter 5: Government

In Chapter 5, Robert Fantina looks at the federal government's management during the Trump administration and its effects. Americans have never known what it would be like to have large segments of the federal government rendered inoperable. The Trump administration was a quasi-experiment in not having a functioning government. Trump's cabinet consisted of individuals from the top one percent of Americans who dominate the nation's wealth and are the only class that benefits financially from racism. Top one-percenters controlled key federal departments—education, environment, and state. One-percenters are already overwhelmingly represented in the Senate, where most Republican senators intend to neutralize the federal government. Under Republican leadership during Trump's administration, no substantial bills benefitting the general public were passed or reforms made. The outcome was the diminishing of virtually all consumer, environmental, educational, and financial regulations and services. Instead, there were predictable tax breaks for the wealthy. The Justice Department lost its independence, and its effectiveness was seriously eroded. Fantina lists some other specific adverse outcomes for all Americans where Whites are the majority.

1. Americans pay the highest prices for pharmaceutical drugs and are forbidden from purchasing high-priced medications from other countries with much lower prices.
2. Under conservative pressure, States have withdrawn funding of state public higher education, which educates 70% of American college students. The consequences are raising tuition costs for parents and ballooning student educational debt, pricing many out of pursuing higher education and advanced degrees. Most of the students harmed by this trend are White.
3. In the largest and fastest-growing states, Hispanics, Blacks, and other children of color are now more than half their age-specific population.

Racial resentment is a factor in preventing federal and state centralized funding of K-12 education. This could eliminate racial differences in school funding, which leads to unequal outcomes. Instead, White voters have been led to believe that local funding of K-12 is preferred. Local funding allows wealthier and Whiter cities and counties to raise more money for their schools while poorer racially-diverse cities and counties struggle to pay for theirs. Consequently, White students have a built-in advantage over others, driving societal-wide inequality, and all American students continue to fall behind the world community in educational attainment.

4. Finally, poverty in the United States has always been portrayed intentionally by conservatives as a Black fault. The government's effort to reduce poverty is a give-away, crutch, and way to advance Blacks and others ahead of Whites. While the American poor are disproportionately people of color, the poor in the United States are mostly White. So, racially motivated opposition to government poverty programs hurt far more Whites than Blacks and others.

Chapter 6: Foreign Affairs

It may seem odd to explore White motives for racism and its effects on White Americans in a chapter on U.S. foreign policy. Not so. The separation of domestic race relations from foreign affairs is a product of the Cold War. In the 1950s McCarthy era, scholars and activists who connected U.S. race relations with international affairs were persecuted by the U.S. government as communist agents or sympathizers. So, the international work of W.E.B. Du Bois, Paul Robeson, Eric Williams, Oliver Cox, and William Patterson, to name a few, were subsequently ignored and forgotten.

The fact is: U.S. domestic racism directly and profoundly influences the conduct of U.S. foreign affairs. In the first and second editions of *Impacts*, Philip White and Gerald Horne reviewed race in foreign affairs. They pointed out that, as a British colony, the United States started as an outpost in British expansion of capitalism and colonization. From the colonies, the enslavement of Africans in the Americas and Caribbean financed European colonial expansion and the beginnings of the industrial revolution. Until the French lost the 1763 French and Indian War, Native Americans and enslaved Africans were potential allies against the British and slavery. Haiti's 1804 violent liberation from the French posed a direct threat to American slavery. In the War of 1812, some Native Americans allied again with the British, and enslaved Africans fled with the British. British abolition of slavery after 1834 threatened American slavery once again. From U.S. Independence until the 1860–1865 Civil War, U.S. foreign policy toward Great Britain, France, and Haiti was defined by the threats these countries posed to a small class of plantation slavers who dominated the U.S. government. Meanwhile, slavery impoverished most White Southerners. Few non-slave owning Whites had access to capital to purchase slaves and rise financially and socially in Southern society.

White and Horne's histories exposed the extent that U.S. foreign affairs defended plantation upper-class's economic interests until the Civil War. It took the Civil War to break their hold and end slavery at a tremendous cost to ordinary Whites. Slaveholders convinced poorer Whites to defend a way of life that only impoverished them. The War cost White Americans 580,000–710,000 dead on both sides. However, the long-lasting outcome of the War was the replacement of one elite, Southern planters, with another, Northern Industrialists. Wealth would no longer be generated by slave labor but rather by wage labor. After the War, emancipated Blacks had the potential to dominate labor in critical industries and, of course, to make wage demands. Blacks knew how to use the vote to reduce economic exploitation after experiencing Black Reconstruction. So, European immigrants were used to offset this Black potential. After all, the Civil War only ended slavery, not racism. Industrialists learned to use Blacks as a reserve labor force to divide labor racially, and produced conflict between Blacks and immigrants over jobs and housing. With Blacks as a reserve labor force, White labor could not demand higher wages, thereby improving owners' profits.

The consistent object of U.S. foreign policy since the Civil War has been to make the national interests the same as national elites' interests. Ironically, elite management of domestic racial hierarchy made the United States a reluctant colonialist. The United States refused to incorporate as States the Philippines, Cuba, and Puerto Rico after the Spanish-American War for fear of increasing rather than decreasing people of color as workers and voters. To this day, American citizens in Puerto Rico, American Samoa, Guam, Northern Mariana Islands, and the U.S. Virgin Islands do not have representatives who may vote in Congress.

Elite control of foreign affairs and domestic racism has affected all Americans. In this edition of *Impacts*, Chris Danielson added to this history and the list of foreign affairs-related effects racism has had on non-elite White Americans. Some of his points are:

1. Racism in early foreign policies helped create a national identity based upon race and White racial superiority. Only Whites could be American, and being White was superior to all others no matter how poor one was.

2. Early U.S. foreign policy helped to make poorer Whites willing foot soldiers of elite White economic interests. Poor Whites put down rebellions by slaves, and other poor Whites exterminated Native Americans, pushed them off their lands, and fought in U.S. overseas adventures.
3. An underlying cause of World War I and II was competition among the U.S. and European colonial nations, plus Japan, for overseas colonies in Africa, the Middle East, Asia, and the Caribbean. Racism was and remains intrinsic to colonialism. The White poor and working-class paid the heaviest price in deaths and casualties in both wars.
4. Southern legislators responsible for racist U.S. foreign policies are also proponents of domestic policies that maintain poverty in the United States and retard modern social welfare development in the United States. To this day, these legislators have consistently opposed improvements in healthcare, education, wages, environmental quality, workers, and voting rights for fear that African Americans and others might benefit. The same people oppose and question necessary measures to prevent COVID-19 infections and deaths.
5. While Donald Trump was President, he and his political base were hostile to globalism, which is increasingly the basis of U.S. economic growth and the American middle-class's continuity. Their anti-global stances have impoverishing another generation of American farmers, threatened American overseas markets and joint-business ventures to the detriment of all Americans.
6. Under Trump, the wealthiest and top one percent of White Americans continue to accumulate larger shares of the national wealth while the working- and middle-shares continue to decline. Sizeable portions of the White working and middle-class support the very policies that impoverish them out of fear of White racial losses.
7. White Southern anti-big government beliefs exist historically to oppose equal rights and economic opportunities for African Americans. These beliefs were used under Trump to do two things. The first was to immobilize the Federal government. The second was to eliminate civil rights and social welfare policies that would have reduced economic inequality and benefitted the majority of all Americans.

Chapter 7: Women

There is a very close connection in American culture between racism and women's subordination. Historically, women are regarded as inferior to men as people of color are seen as inferior to Whites. In this edition of *Impacts*,

Charlotte Dunham uses "controlling images" to explore the effects of cultural racism on White women. Controlling images are narratives about groups and individuals that explain "who they are" in a way that supports the gendered and racial social order as being natural and inevitable. Dunham identifies how several of these images are used to control White women and showed their effects:

1. The Trump administration used racist controlling images of poor Black women to propose cutting food stamps (SNAP) and other programs during the COVID-19 pandemic. Far more White women and their families would have been harmed by these cuts than women of color and their families.
2. The use of racist controlling images of poor Black women negatively impact White women's sense of identity. Some White women qualify for welfare but refuse to take it because needing assistance would diminish their status as Whites and equate them with Blacks.
3. Blacks, Latinx, and Native peoples have proportionately more women in poverty and need more essential assistance than White women. However, there are numerically many more White women in poverty who need assistance. Anti-welfare and assistance sentiment ends up hurting many more White women and their families. They, too, must participate in inadequate programs based on false premises about poor (and Black) women's character.
4. There is a heroin epidemic sweeping through White U.S. suburbs and rural American communities. Criminal justice penalties designed to punish Black women who abuse and deal drugs are now being applied to many more White abusers and dealers. White privilege is not enough to protect poor White women from punitive laws intended to punish poor Black women.
5. Most nursing home residents are White women. The number of homes will increase for the foreseeable future. Immigrants from El Salvador, Guatemala, Nicaragua, and African countries are the most likely source of attendants. These are precisely the people that Trump and Conservatives wish to keep out of the United States with border walls, separating children from their parents, and ending asylum programs. Harm will come to elderly Whites due to a lack of staff.
6. Racist controlling images have made it difficult for White women to engage with women of color as peers in work for women's rights and equity. These women must first recognize that they hold racist stereotypes and have pretenses before they can work effectively with others. This issue is challenging for White middle-class women to acknowledge and address.

Part II: Individual and Case Studies

Cultural racism does not determine outcomes. Fortunately, social and personal circumstances moderate how individuals react to cultural messages, as we saw in Spanierman and Clark's chapter. We summarize Brooks, et al., study of a student-athlete boycott, and Patricia Maloney's study of racial transitions in teacher attitudes. The following are reviews of the chapters in Part II.

Chapter 8: Social Psychology

In the first edition of *Impacts*, Thomas Pettigrew (1981) pointed out that not all White Americans accept racism. Based on decades of survey research findings, approximately 20% of White respondents consistently oppose racist propositions, whereas another 20% consistently support racism. The remaining 60% and the majority appear to have no compelling motivation to accept or reject White supremacy. They go along with prevailing institutional and cultural practices. That is: they believe and do whatever is current wherever they are.

In the second edition, Jones and Carter sought to investigate Pettigrew's hypothesis. For European Americans whose White identity is assumed, unquestioned, and taken for granted, the historic cultural script of White supremacy and non-White inferiority is a central part of their self-identity. Like anything that is unquestioned and deeply held, this identity and its racism are defended vigorously. Whites who have had no exposure as peers with people of color and have never had the opportunity to wonder, question, or explore their racial and cultural identity are what Jones and Carter refer to as in the "contact stage" of White identity. They are likely to be Pettigrew's racist 20% of Whites.

Pettigrew's 60% who simply conform are likely to be in what Jones and Carter call the "disintegration and reintegration" stages of White identity. They, too, have absorbed the historical culture of White racism but are in varying degrees aware that something is inconsistent between what they have experienced and what they believe about race. Their encounters with people of color have not led them to question their assumption of supremacy, but they have some doubts about what they have been told and what is idealized about others. They are influenced easily by racial stereotypes and shifting conventions in general opinion about race. They do what anyone would do under the circumstances: try not to think about it and go along with acceptable racial convention. So, most White Americans can support civil rights measures in one decade and be influenced by racial symbolism and innuendos through political campaigns in the next decade. Then, the same people can be convinced that racism does not exist and that instead, they are the ones facing racial discrimination.

The 20% of Whites in Jones and Carter's final stages are more likely to be anti-racists. These are people whose life experiences have led them to look inside themselves and question White racial supremacy. Working with people of color as peers has led them to shatter racial stereotypes in the historic scripts that they initially held.

In this edition, Lisa B. Spanierman (a former student of Robert Carter) and D. Anthony Clark further ground Jones and Carter's work empirically. The psychosocial costs of racism to Whites (PCRW) theory demonstrates that White Americans face negative cognitive, affective, and behavioral consequences of racism because they are in the dominant societal position. The PCRW scale measures the affective costs of racism to Whites. We learn that White empathy, guilt, and fear are meaningful aspects of White people's racial experiences. And when we consider White empathy, guilt, and fear in combination, we know that five different types of White Americans emerge. These types range from blatantly racist and insensitive to those who are empathic and anti-racist.

Spanierman and Clark write, "White racial affect is likely both a curse and promise in the age of Trump. As a curse, it is foundational to reproducing material inequities, as insensitive and afraid types report Black people to the police for innocuous activity, thereby potentially subjecting them to police violence. As promised, White racial affect offers the possibility that White people can become part of the solution rather than remain complicit with racism. Being part of the solution is indicated by the unprecedented number of anti-racist Whites who have joined the protests against police violence in the wake of the deaths of George Floyd, Breonna Taylor, and others."

White identity's role in deriving racial orientations among White Americans picks up two other related insights from the first edition. Rutledge Dennis (1981) described the psychological fallout from racist socialization. It is not straightforward. White children are sometimes confused and even traumatized in learning to regard and treat White people one way (superior, good, and safe) and Blacks and people of color another way (inferior, bad, and dangerous). Robert Terry (1981) described for White adults the effect of learning to dehumanize people based on race. He points out that racism requires one to be inauthentic. That is to hold religious, professional, and even personal beliefs that are entirely contradictory to one's prejudicial racial beliefs. This leads an individual to say one thing and do another and hide the contradiction from others and oneself. Even worst, it means socializing the next generation to do the same—live in denial of their racial and moral contradictions, and never address and remain unconscious of their dilemma.

Chapter 9: Media

Several developments play a critical and predictable role in the development of this nation's conservative-liberal polarization. In the 1950s, everyone watched the same television shows and heard the same news, whether it was from ABC, NBC, or CBS. The evolution of national networks from broadcast to cable television fragmented the common audience. This fragmentation produced the potential for each audience to receive completely different news and commentary. Today, while cable television's entertainment content is virtually identical across cable networks, news content and commentary are not. The intentional creation of a conservative media began in earnest with the Australian Rupert Murdock's entry into the American market and his discovery that news with conservative political commentary is profitable. Then, MSNBC was started to provide a counter commentary, while most other news organizations attempted to remain politically neutral. The outcome is not just audience fragmentation by political liberal-conservative points of view, but polarization. The advent of social media has only multiplied the fragmentation where now micro-audiences see and hear only what they want to see and hear and live in isolated realities.

In this edition, Kalimba Kizito looked critically at Fox and other news organizations. What Kizito reveals is that Fox News and its audience are distinct from the others. Its most popular programs are not news but commentaries by Sean Hannity and Tucker Carlson, which, by the way, do not make any claim to be accurate and fact-checked. A large segment of the Fox audience relies only on Fox for their news and commentary. All the other news networks, including MSNBC, report extensively on multiple sources and fact-check their news and commentary. High proportions of non-Fox News audiences also use multiple sources for their news and commentary.

In effect, Fox News operates in an information silo where its audience misses most domestic news and virtually all international news. Additionally, there is little distinction between news and commentary. According to Kizito, Fox "dances to its own tunes." Consequently, the Fox News and commentary audiences are the least informed, the least knowledgeable, and the most misinformed of news audiences. The fact that Fox News became a virtual extension of Donald Trump's midnight tweets only deepened their silo, made the echo louder, and, over four years, made the most loyal Fox viewers more and more misinformed, angry, resentful, and openly racist. Ironically, behind the scenes, until 2019, Rupert Murdock and other Fox News executives were very concerned about their isolation, draft, and some of the "talking heads" lack of professionalism. But their concern was not enough to change their news formula. Whatever they are doing is very profitable, and they are in the news business to make money.

The outcome is predictable. A significant proportion of White America, who constitute the Fox audience, has been led into a worldview where they believe their freedoms and government have been taken from them. Blacks and other people of color are the foot soldiers of communists and socialists. Immigrants, rapists, and criminals are overrunning the country, and White people's days are numbered if they do not act. The Fox News audience is thoroughly alienated from the federal government, has radicalized the Republican Party, and is open to bizarre conspiracy theories. A high proportion, if not the entire mob that attacked the Capitol and attempted to take over the U.S. government on January 6, 2021, consists of loyal Fox viewers. Viewers of no other mainstream network have such extreme views and have worked themselves into such a close-ended nightmare. Fox News and commentary have hard-wired many White Americans to cultural racism and have propagated racism for another generation. The racism that is a part of Fox-driven alienation is clearly harmful to White Americans.

Chapter 10: Sports

A chapter on racism and sports is new to *Impacts*. College and professional football and basketball are areas in American life where there is supposedly a Black–White win–win. Black athletes gain fame and fortune, and White fans can enjoy extraordinary athletic contests. Scott Brooks and colleagues provide a case study of the racism underlining this win–win, which illustrates how cultural racism is deployed and its corresponding effects on Whites. It turns out that the Black student-athlete role is highly restricted; they are at the university only to play their game. They are expected to be grateful for their good treatment and not express interest in social or racial justice issues. This is the opposite of what is expected of other students—to think critically and be civically engaged. Under no circumstance are they to make playing their games leverage for anything. If they do, they are punished like striking industrial workers, not students.

Brooks and colleagues show not only what happens to Black athletes who get out of line, but they also showed what happens to White staff and faculty who side with the Black athletes or who get in the way of their punishment. Brooks et al. write, "The handling of Black protests by the university's administration and board of regents/trustees shows their investment in racism despite the potential damage inflicted individually and collectively on Whites involved—the university's president, chancellor, athletic director, football coach, student-athletes, and general student population. The reputational stench has created a loss in applications for admission and a crumbling of the flagship campus." Suppressing a Black protest was costly to Whites as well.

Chapter 11: Education

Then, Patricia Maloney, in Chapter 11, shows the subtle impact of racism on White school teachers. White teachers, especially those in less supportive and resource-poor schools, lost their initial sense of empathy for their students and their critical self-reflection. They have less job autonomy, felt more alienated from their work and students, reported less job satisfaction, experienced more burnout, and had higher turnover than White teaches in a more supportive environment. Ironically, they begin their careers enthusiastic and empathic toward their minority students but end up indifferent and even hostile five years later. They were led to believe that they were teachers and educators. Instead, they found themselves in institutions whose primary mission was housing and controlling students of color to replicate racial hierarchy. Education was an institutional pretense. The students and parents knew this and treated the teachers accordingly. The students were oppressed, but so were the White teachers.

Part III: Anti-Racism and Conclusion

Chapter 12: Social Movements

In Chapter 12, Pamela Oliver provides a big picture to understand White American's unprecedented involvement in recent #Black Lives Matter demonstrations. She shows that the struggle for racial equality in the United States has never been Black versus White or a straight line of victories or defeats. A careful look reveals that Whites have been on both sides of the struggle against slavery and Jim Crow, and racism today. They have never been united in oppressing people of color or in enjoying White racial privilege. In reviewing the history of social movements, we should not be surprised that there is active White support for racial justice today. What is unprecedented is the scale of that support. This realization has profound implications for how we view social change and the likelihood that a reduction, if not elimination of racial inequality is possible.

Chapter 13: Affirmative Action

In Chapter 13, Fred Pincus explains what affirmative action is. He dispels the conservative truism that affirmative action is about reverse discrimination against White Americans. Affirmative action has benefited everyone who has ever applied for work or applied to a college or university because it establishes a procedure for fairness and makes decisions less arbitrary. Intentional interventions into the racial practices and outcomes of institutions are the only effective way to correct historic and contemporary racial inequalities. New affirmative action-type measures will have to be taken to reduce inequality further.

Implications

We should not overlook consistent points in contributors' chapters. The first is that the vast majority of European Americans are not the source of a belief in White supremacy. Racism does not spring from them without prompting; it has to be propagated to them as is the case with Fox News, Rush Limbaugh, and racists on social media. The second point is that Whites in positions of influence, especially political leaders and media commentators in this age of Trump, are the primary source of racism's propagation. The people who propagate racism do so to gain and keep control over institutions for personal gain and their elite benefactors' gain. Finally, European Americans are harmed as well as people of color by the propagation of racism and the use of institutions to maintain White advantage and privilege.

Finally, in this book's introduction, we identified two perspectives on White identity and racism. In the first, racism is intrinsic to White identity. White identity and racism cannot be separated. The only way to eliminate racism is to do away with White identity. In the second perspective, White identity politics, White identity and racism are distinct entities. One exists without the other. Which perspective is correct?

Upon reviewing four decades of scholarship on White identity and whiteness, we conclude that there is only one acceptable type of White identity. In this third volume of *Impacts*, Spanierman and Clark describe five White racial orientations: insensitive and afraid, oblivious (unempatic and unaware), empathic but unaccountable, fearful guilt, and anti-racist. Only the anti-racist White identity can work to dismantle racism, increase racial equity, and support social justice. An anti-racist White identity must deconstruct Whiteness itself. This means working through the ways in which colonialism, conquest, privilege, and advantage have shaped and formed Whiteness. The anti-racist White identity must also grapple with social and psychological costs that Whites suffer from supporting White racism.

Without a purposeful anti-racist level of introspection, all other forms of White identity are prone to political and social manipulation for the benefit of White elites. Donald Trump stoked the flames of White shame, White guilt, and White fear to build political power and to cling to that power when he was voted out of office. At no point during his time in office did working-class Whites benefit materially; they only benefitted symbolically. Instead, billionaires received tax breaks and convicted associates received pardons. Meanwhile, the mostly White supporters who descended into violence during the January 6, 2021 insurrection at the nation's capital received criminal charges.

In our concluding chapter, we review what we think can be done to alleviate White racism and the challenge before us. We explain what is at stake and what we might anticipate in the decade to come.

Discussion Questions

Q1: Name 3 ways that White Americans are impacted by racism.

Q2: Name 3 ways that Donald Trump stoked racism to benefit himself or other wealthy elites.

Q3: Name 3 ways that White Americans can support anti-racists movements or become anti-racists themselves.

References

Chetty, Raj, Nathaniel Hendren, Patrick Kline, and Emmanuel Saez. 2014. "Where Is the Land of Opportunity? The Geography of Intergenerational Mobility in the United States." *The Quarterly Journal of Economics* 129 (4): 1553–623.

Estep, Kevin and Pierce Greenberg. 2020. "Opting Out: Individualism and Vaccine Refusal in Pockets of Socioeconomic Homogeneity." *American Sociological Review*.

Knowles, Elizabeth M., ed. 1999. *The Oxford Dictionary of Quotations*. Oxford: Oxford University Press.

Li, Huiping, Harrison Campbell, and Steven Fernandez. 2013. "Residential Segregation, Spatial Mismatch and Economic Growth Across US Metropolitan Areas." *Urban Studies* 50 (13): 2642–60.

Samila, Sampsa, and Olav Sorenson. 2017. "Community and Capital in Entrepreneurship and Economic Growth." *American Sociological Review* 82 (4): 770–95.

15

Conclusion and Reflections: Impacts of Racism in the Age of Trump

Benjamin P. Bowser and Duke W. Austin

Abstract In this, the third edition of Impacts of Racism, we reiterate the main points made in 1981 and 1996, and then build upon and extend them with new findings, insights, and prescriptions. Previous editions pointed out that if White racism was not sufficiently addressed in the ensuing years, our national circumstances would become more desperate. The violent attack by White American "patriots" on the U.S. Capitol on January 6, 2021, showed precisely that. Racism was intrinsic to the attack as it is to the continuing danger posed by the attackers and their millions of sympathizers. January 6th told us a lot—the most important of which is that we do not have another forty years to ignore racism and do nothing about it. In this concluding chapter, we discuss strategies to disarm and dismantle racism on cultural, institutional, and individual levels.

Keywords January 6th insurrection · Cultural racism · Institutional racism · Individual racism · Solutions · Policy solutions

B. P. Bowser · D. W. Austin (✉)
Department of Sociology, California State University, East Bay, Hayward, CA, USA
e-mail: duke.austin@csueastbay.edu

B. P. Bowser
e-mail: benjamin.bowser@csueastbay.edu

D. W. Austin and B. P. Bowser (eds.), *Impacts of Racism on White Americans In the Age of Trump*, https://doi.org/10.1007/978-3-030-75232-3_15

It is worth repeating two of our central points in this anthology. First, African, Native, and other people of color are not the source of White Americans' fear and need to subordinate them. "The race problem" does not exist today because of something inherent in people of color nor because of their behaviors. Racism exists because White people in power use it to unite European Americans so that they can continue to gain wealth and stay in power unincumbered. They need racism against people of color to distract from their exploitation of vastly more White Americans. Second, we have shown that people of color are not the only ones negatively impacted by racism. White Americans pay a heavy price as well for maintaining racial inequality and White elite's special privileges.

The implications of these two points are straightforward. The main burden of reducing, if not eliminating, racism falls on White Americans. Since a relatively small group of Whites in each generation are the source of the problem, other White Americans have the primary responsibility to confront them and stop their propagation of racism. White Americans have the human and material resources to directly oppose appeals to Whiteness and the racial fears that White politicians and billionaires use to win elections, sell products, and maintain advantages over everyone else. People of color do not have nearly the same resources to directly challenge White power.

We challenge the assumption that White material and status advantages motivate most European Americans to continue racial subordination of non-Whites. Not only is the source of racism's propagation specific to a small class of Whites, the cost of racism to White Americans is not evenly spread. All social classes of White Americans pay a psychological and spiritual price while the economic costs are particularly high for the lower and middle classes. We also challenge the assumption that American society and its economy are zero-sum games, where one race's gain is another's loss. There is enough growth in a modern, dynamic economy for all talent and effort to be rewarded based on achievement, not race. As Michael Reich explained, the economic benefits from racial inequality get progressively higher as one goes up the income range to the top 1%. In contrast, White workers in the lower to middle-income ranges are impacted negatively by racism and have a strong economic reason to oppose and not support it. Among elites in the United States, conservative and libertarians are the only ones who do not understand that they can continue to gain wealth and maintain power without impoverishing other Americans in a zero-sum game. Several chapters in this anthology show that non-elite Whites are impoverished and physically and materially damaged by racism. Both the obvious and not so apparent costs of racism for White Americans are summarized in the previous chapter.

Urgent Times

This is the third time in 40 years that a series of investigators have looked in-depth at White racism, its motives, and what impacts it has on White Americans. The core of what each team presented 40 and 25 years ago, respectively, is still timely. This third edition reiterates the main points made in 1981 and 1996 and then builds upon and extends them with new findings, insights, and prescriptions. In the second edition, it was pointed out that if White racism was not sufficiently addressed in the ensuing years, our national circumstances would become more desperate. The violent attack by White American "patriots" on the U.S. Capitol on January 6, 2021, showed precisely that. After overwhelming police guards, killing one, and injuring over one hundred, they searched the building to capture, injure, and kill lawmakers, including the Speaker of the House and the Vice President. January 6th is an impact of racism on White Americans. Racism was as intrinsic to the attack as it is to the continuing danger posed by the attackers and their millions of sympathizers. January 6th told us a lot, the most important of which is that we do not have another forty years to ignore racism and do nothing about it. What were the messages the attack gave us about racism, its impacts, and the future?

The first message is that the attackers are no longer on the fringes of American society. White nationalists, Neo-Nazis, members of the Alt-right, Neo-Confederates, White conspiracy theorists, and their sympathizers are to be reckoned with now. All are self-avowed racists. None of these groups use indirection in talking about race, nor do they sugarcoat their antipathy for non-Whites or Whites who do not share their racial antipathy. Furthermore, they are now embedded in one of the nation's two major political parties. They have the sympathy and support of Republican elected officials who excuse, minimize, and even defend their actions.

The second message is that the misinformation that led to their extreme beliefs and attack on the Capitol has not stopped. Former President Trump's angry speech before the attack only inflamed passions and beliefs already there. Trump claimed he won the 2020 national election and that the Presidency was about to be taken from him only because of voter fraud in Black electoral districts—a claim he nor anyone else could prove after fifty lawsuits. He urged the crowd to go to the Capitol and stop Joseph Biden from being certified as President. Most explanations of the January 6 attack fail to account for the misinformation spread by Fox News commentators, commentators who have the three most popular shows on cable television. These commentators have spent years providing their predominantly White audiences with dire warnings, us-versus-them, stoking fear, providing misinformation, and dehumanizing narratives about anything and everyone unfamiliar to them. They openly advance and defend White superiority as the same as Americanism. According to Fox News, America is and has always been perfect, and Indian genocide did not happen. Africans' enslavement was benevolent, and any idea, fact, or criticism that questions American exceptionalism is anti-American and communist-inspired. The problem is that this xenophobic and paranoid worldview continues after the attack and is defended as Fox News' first Amendment right to free speech. The consequence is that millions of Americans fully believe their racist and error-filled narrative and support extremists who now dominate the Republican Party.

The third message is that the attackers' grievances are not primarily economic. Most were well employed, could afford to fly to Washington, and were able to stay at hotels; they were a cross-section of working and middle-class Americans. More jobs and economic growth will not cool this fire. White identity politics drive their conservatism and grievances. They believe they are losing their control of the country, losing racial privileges to Blacks and immigrants, and have to fight to keep what is "theirs." This means maintaining White (Republican) dominance of political offices and government supersedes traditional respect for the will of voters. This is especially the case now that a substantial proportion of voters aligned against them are people of color. Therefore, the January 6th rioters felt justified and, indeed, that it was their patriotic duty to stop certification of a "fraudulently" elected President. They wanted Donald Trump to remain in office even though he lost the election and there was no evidence of voter fraud. Impeaching Trump and prosecuting rioters has not stopped Fox News or its successors from stoking this White supremacist fire. So, from here on, after 2021, elections in the United States are going to be mortal combats between those who wish to continue a democratic tradition of government and those who are willing to sacrifice it if they do not win and elections.

What Can Be Done? Negate Cultural Racism

The accumulative view from this and two prior editions of *Impacts* is that cultural racism must be negated at its source. Unless we reject cultural racism, it will continue to fuel institutional and individual racism. Without eliminating all three, discrimination and prejudice will persist, as will the impacts of racism on Whites.

As we discussed in Chapter 1, *White identity* describes the degree to which White people identify with *Whiteness*, the idea of the White race. Whiteness, however, was founded on the ideology of White supremacy, and is therefore a form of cultural racism. Must we, therefore, reject Whiteness completely in order to establish a more just and equitable society?

Certainly, some scholars and activists have advocated for the elimination of White culture completely, but we take a more nuanced view. In the first edition of *Impacts*, Thomas Pettigrew (1981) pointed out that not all White Americans accept racism. Based on decades of survey research findings, approximately 20% of White respondents consistently oppose racist propositions. In the second edition, Jones and Carter (1996) confirmed Pettigrew's findings and identified a minority of White people whose life experiences led them to look inside themselves and question White racial supremacy. Working with people of color as peers led them to shatter racial stereotypes in the historic scripts that they initially held. Then in this edition, Spanierman and Clark (2021) identify five racial affect types of White people based on combinations of White empathy, fear, and guilt. The final racial affect, *anti-racist*, is consistent with Pettigrew's, Jones', and Carter's previous work.

We, therefore, recognize that Whiteness, aka White culture, is founded in and in many ways is synonymous with the ideology White supremacy. However, there exists a subset of White identity that is introspective and rejects White supremacy. That subset is the anti-racist White. White people who wish to reduce or eliminate racism must reject the part of Whiteness that is deeply tied to White supremacy. They must overcome the insensitivity, obliviousness, and fear that dominates Whiteness, and they must increase their empathy for each other and people of color. As anti-racists, they must grapple with the origins of Whiteness and of the White race itself. This means acknowledging and confronting their White privilege, and increasing their multicultural education. It would also be helpful to intentionally diversify their social life, building close friendships across racial differences, and cultivate respect and support for their professional colleagues of color. An anti-supremist would support public policies, such as affirmative action, that endeavor to counter institutional racism, and confront instances of individual racism. In short, White people must reject White supremacy and become anti-racists.

Negate Institutional Racism

What if we could successfully eliminate institutional racism in employment, and no other institutions were changed? We keep everything else the same except in employment. Would that be the end of racism in employment? We know from experience the answer is "no." Such an achievement would be met by elite orchestrated blowback through other institutions. There would be lawsuits (courts), state propositions outlawing the change (state government), opposing media like Fox News, and politicians running for office against the changes. That is not all. The general belief that since Whites are superior, they should have better outcomes in employment will be used to discredit the changes. Eliminating White advantage will end up as discrimination against Whites, an attack on White supremacy, and violation of long-term norms. It is unlikely reforms in employment could survive such an onslaught for very long. This is precisely what happened to affirmative action (see Pincus). Eliminating racism in employment could not be sustained without eliminating racism from other institutions and from generalized beliefs as well. So, to undercut institutional racism would require doing so across multiple institutions. The change would have to be systematic and broad based. The following suggestions illustrate how systemic change to eliminate institutional racism and its impacts will have to be approached in this age of Trump:

Control of Government: Control of federal and state governments is necessary to maintain institutional racism. The same control of government is necessary to weaken, if not eliminate, institutional racism. Government is the only institution large enough to make a systematic change that has mandate to work for "the common good." This is why contestation over control of the federal government is so crucial and is the basis of the current national crisis and urgent times. The recent effort to eliminate democracy for right-wing autocratic control of the federal government would, for the first time in American history, negate the federal government's potential to address racism and a host of other systemic problems.

The only way to beat the Trump assault on government without a new Civil War is for a large and strong enough coalition to form around preserving democracy (labor, Democrats, anti-Trump Republicans, racial-ethnic groups, women, and anti-racists). Advancing anti-racism and eliminating the adverse effects of racism on Europeans, Africans, and all other Americans require this coalition to vote en masse in future elections. They will have to do so as long as it takes to politically defeat those who believe that the federal government's control is their racial birthright. If they cannot control the government, they will destroy its potential to serve any other interests than their own.

Elections: If government control can be maintained long enough, a practical way to reduce racism and its adverse effects in politics is to impose spending limits on what candidates and campaigns can spend on elections. What fuels racism is money, lots of it. Wealthy individuals and their interest groups should be identified as well. This will limit the extent that elite dark money and deep pockets can distort elections and overwhelm voters with misinformation.

There should be an independent national commission devoted to monitoring election district boundaries. States should apply to the commission to approve changes in the boundaries of any election district. The change cannot go into effect until the state receives commission approval. The commission's job is to determine if the proposed change is needed, necessary, and fair.

Online registration and voting with verification would also strengthen the democratic process and advance anti-racism. "My Vote" online would be another alternative to mail-in and in-person early and same-day voting.

Education: Teaching about racial and cultural diversity and valuing it are not enough. There is a need for anti-racism to be integrated into K-12 and higher education curriculums. Cultural racism is passed across generations because it has never been acknowledged and explained widely, except in a few college and university courses. Even those who acknowledge racism's existence lack enough knowledge about it to explain racism, its origins, and how it affects them. They cannot explain racism to their children. Anti-racism education is not Black and Chicano history for Blacks and Chicanos, although this is important; it also teaches European American students about racism and Whiteness. Furthermore, such an education does not require the propagation of fiction, morality lessons, or political correctness. There is now sufficient knowledge at the university level in all the academic disciplines to devise credible anti-racism curriculums for K-12 and public education.

The age of Trump has revealed another educational deficiency, general knowledge of civics. Former President Trump revealed a shocking ignorance of the separation of powers, the legislature's role, separation of church and state, and federal departments' organization and functioning. His ignorance is shared by a large portion of the American public of all political persuasions. It is ironic but fortunate that so many of those who attacked Congress on

January 6, 2021, failed to capture legislators because they lacked knowledge of the historical building and had never been in it before. So, civics or general education about the organization and function of a democratic government must be reinstated in our schools. Political cynicism, anger, and indifference to the government are likely outcomes of not knowing how government works. Also, one can be easily misinformed and manipulated by lacking basic knowledge of civics.

Economy: Michael Reich has shown the intrinsic relationship between the economy and racism and their impacts on workers and nonworkers of all races. Virtually any economic measure that reduces social class inequality is also an anti-racism measure. When lower and middle-class workers' wages and benefits improve and profits are moderated, racism is reduced. Spreading income and wealth more equitably down the social class hierarchy will reduce racism and reduce racism's negate impacts on all Americans. A minimum wage tied to the cost of living, a national health plan, and enhanced social security and unemployment insurance are several other measures that can reduce racism and its harmful effects.

Social Media: In the age of Trump, there is open, vial, and unchecked racism in social media. Racists and their sympathizers gang up on and attack non-sympathizers, form their own closed world, and reinforce their beliefs, which become alternative realities and the basis of bizarre conspiracy theories. Racists participate anonymously and hide their identities behind avatars and online personas, but are more than willing to discover others' actual names and addresses in order to threaten and intimidate them. This is a direct negative effect of racism on European Americans. Such bullying can be done anonymously and with no consequences—a perfect environment for socio- and psychopaths to thrive. Anti-racists need to form online organizations and teams and then devise strategies to engage racists while protecting themselves. Racists should be challenged, informed, debated, and even outed with their real identities disclosed. Perhaps, none of the avowed racist's minds will be changed, but newcomers and others who read challenging and alternative messages might be given reason to pause. If challenged widely enough, it would be virtually impossible to create a completely closed social space and racist alternative reality.

Broadcast Media: An independent and, perhaps, even bipartisan panel could establish certification standards for what constitutes best practices in broadcast news gathering and reporting. It is much more difficult to propagate racism and its damaging effects through news organizations with high ethical and professional standards than through organizations without standards. The panel can periodically evaluate the extent to which news programs and organizations meet the standards and make the results widely known to news audiences.

More independent fact-check organizations are needed to monitor the accuracy of radio-television news and commentary routinely. Listeners and viewers should see conveniently available fact-check of news and commentary promptly. Besides, we might consider an online Wikipedia of nonsense, fictions presented as facts, misrepresentations, conspiracies, and lies.

Activist Strategy: Government cannot do everything. Anti-racism activism is necessary to call attention to and define social problems that need to be addressed. However, there is a strategic error that activists commit where their hard-fought progress in reducing racial inequality is easily lost. The error is to narrowly focus on reducing racism or discrimination in housing or education or employment or the media or criminal justice. Successes in one-issue reforms are wiped out easily. More effective anti-racism will require attacking racism on multiple fronts simultaneously. Any advances will be better sustained if racial inequalities are reduced across several institutions at once. As a matter of practical necessity, multi-issue coalitions will have to work together to execute such broad-based strategies. Soon after any gains or successes, planning will have to begin devising counters to inevitable efforts to undo the change.

In our view, lawsuits are an under-utilized way to fight racism and its effects. Racists thrive on pretense—historical fictions, racial stereotypes, and defaming individuals and organizations. Donald Trump's Twitter content provided a highly visible example of such racist pretense. What would happen if Fox News anchors or commentators were sued every time they uttered a falsehood about someone or misled the public—such as opposing mask-wearing in the middle of a pandemic? They are liable. If they lost enough lawsuits, their networks would have to weigh the consequences of further defamation. Furthermore, continuous suits and a few wins would threaten the organization's financial bottom-line and their reputation. Racism would become expensive.

Negating Individual Racism

Spanierman and Clark have provided empirically grounded variations in European American racial dispositions. Not enough is known in how those who are most inclined to racism can be transformed into becoming less inclined and perhaps even anti-racists. The key to such a transformation for most is some yet to be determined combination of negating cultural racism and participation in institutions that do not reinforce racism. A way to do both on a more modest scale is through individual and group counseling.

At the individual level, pointing out and questioning beliefs in White racial supremacy and fear of racial others should be integrated into psychotherapy and counseling for White Americans. There is an additional possibility. European Americans strongly attached emotionally to a White identity may have intergenerationally unresolved historical grief and trauma. Their parents or grandparents may have brought these issues to the United States as immigrants. Or the trauma may have happened since their arrival. Loss of ethnicity may have been traumatic in itself and may become evident only when their Whiteness is challenged.

The American Psychiatric Association's (APA) failure to address racism and the psychology of White supremacy is one reason we know so little about the effects of racism on European Americans. We can now add White supremacy as another psychologically negative impact of racism. Also, counseling in groups can reach more and may be more effective in helping European Americans deal with racism. As suggested in the first edition of *Impacts*, "if groups can first depersonalize the issue of racism by examining generic principles, such as up/down relationships that we all have experienced, it is easier and safer to personalize and 'own' White racism in one's self as a White person and to communicate with others about it." However, even if the APA Diagnostic and Statistical Manual was revised to include White supremacy and racism, individual and group counseling would be insufficient to eliminate racism in the United States. But it is a start.

A Look Ahead

In the 1981 first edition of *Impacts* we advanced a belief that a serious attempt to negate racism and reduce its impacts on Whites, and certainly on people of color, will require government. Only the federal government could regulate what corporations do to gain profits, prevent them from further impoverishing the general population, limit the power and influence of special interests, and more aggressively protect "the common good." However, national and international economic trends are increasingly made in corporate boardrooms, not in legislatures. It is in the private sector where most Americans are employed (and unemployed), where single corporations are larger in wealth and influence than most Western governments. Their investment decisions can make and break cities and states to say nothing of political careers. Increasingly, the government's role is to react supportively to any private sector initiative and accept responsibility for the social fallout of board room decisions with social services, environmental clean-ups, and law enforcement.

Since the 1996s edition, we have watched coalitions of increasingly conservative interests, facilitated by billionaire libertarians, come together to dismantle and restrict government rather than to make it more effective—"reduce it to the size of a baby in a bathtub and then drown it." Meanwhile, the negative effects of racism continue to accumulate and erode the quality of life and the well-being of everyone. Only the top 1–5% of Americans have gained financially in recent years. Yet everyone, including the top 1–5%, has experienced a continued loss of personal security and optimism, heightened anxieties, and increasing alienation from society, government, and work, along with a general decline in civility and public regard for one another. Between the board rooms and conservatism, we are driving in the wrong direction.

Heading to Dark Places

We were at a crossroads at a dark place in 1996. Now, we are at an even darker place. Racism's centrality in the nation's crisis is not fake news and leaves no room for pretense, to hide, to muddle through, or delay for another decade or two. We are now in the age of Trump, and the patriot assault on the U.S. Capitol on January 6th has made it clear that inaction is not a choice. Our house is on fire! Racists are frightened by the nation's evolving demographics. Minorities of color will soon become the nation's majority. The first large minority cohort of children are now young adults and will soon be able to vote. The Republican Party's continuing use of the "the Southern strategy"—using covert racist appeals to Whites—can now backfire by uniting these new voters against them. Increasingly, conservative White politicians who stayed in office by stoking racial fear among White Americans are vulnerable; they can be voted out of office by disaffected Black, Brown, and White voters. The future of several voter-rich red states is blue. Texas, Georgia, Arizona, and South Carolina. In the racist imagination egged on by Fox News, this is a nightmare and the end of the White race. Retailers are already abandoning them and are adjusting their commercials to reflect a multiracial America. They realize that their future profits are dependent upon Black and Brown customers. Nothing is going to prevent Whites from becoming a demographic minority. So, what are the choices?

Choices

The first choice is the California option. White Americans are already a minority in the State of California. However, the state's White liberal elite is still very much in power, has lost no social or economic privileges, and is not about to be deposed. Liberal White elites have an enlightened way to exercise power. They promote and respect racial diversity and "big tent democracy." They even allow the state legislature to reflect the state's racial diversity. However, they do what elites do best—control the money. Black and Brown legislators, trustees, and board members are just as beholding to elite funders as their White counterparts. Ironically, racial dog-whistles and the Southern strategy no longer work in California due to growing Latinx political power, even in the conservative counties that launched Richard Nixon and Ronald Reagan. The State Republican Party is only a shadow of its past.

In comparison to conservatives' open racism, there is virtually no attention and criticism to liberals' racism in blue states such as California. They also maintain inadequate minority schools, racially segregated housing, and defund higher education because there are too many non-White students. They allow racially abusive policing, and keep prisons full of people of color. Liberal elite racism still waits to be addressed if the fire of in-your-face conservative racism can be contained.

The second choice is the Mississippi option. There is a way to prevent losing control of state and federal governments to maintain economic and social privileges. When White supremacists find themselves as a minority among their alleged inferiors, they do the following. They seize control of government, do whatever they can to disenfranchise the non-White majority, put a security state in place to control non-Whites, and rule with an iron fist through terror and violence. This is precisely what happened after the U.S. Civil War in Southern states with African American population majorities—such as South Carolina, Mississippi, and Alabama. The historical and continuing racism in these states is why they are still today economically the poorest and socially least progressive states in the nation. The same White minority control of the majority population is what South African apartheid was about—modeled after Jim Crow in the U.S. South.

Our two major political parties are now primed to produce these options. The Democratic Party is heading to the California option. If the Trump patriot-base solidifies their control over the Republican Party, the Mississippi option is their inevitable direction. Who will win? As already suggested, the Democratic Party "the big tent" coalition will have to hold together and consistently win state and national elections for the foreseeable future—something they have never done. In contrast, it will not take much for the Republican Party to make the Mississippi option the future. Elect another Trump as President and take control of the Senate once again. Then, do a better job of dismantling the federal government and make it impossible to be voted out of office. They can do this by declaring enough future other-party votes as fraudulent. This scenario is the nightmare we face if Fox News continues, the Trump patriot-base is not repudiated from within the Republican Party, a sizable number of people do not defect from the Party, or a third party fails to form. Democracy will end, and racism in all its forms will go viral.

The dark place we are now in is where Germany was in 1933. In a series of elections shortly afterward, enough Germans were frightened into voting the Nazi Party into office—Jews were the scapegoat. Then, the Party began "Nazification" of the German government and ended free elections. German voters unwittingly supported the end of their democracy. To make America Great and save it for White people, the Trump patriot-base has already shown us their willingness to do the same. What no one anticipated before Trump was how deep racism is engrained. In Chapter 3, the men Jonathan M. Metzl interviewed, who were willing to die of preventable causes to keep Blacks and immigrants from getting health care, indicate just how deep racism runs in the United States. There are elites and a significant number of White Americans who are willing to end American democracy before they let people of color fully participate in it—as a majority. The tragedy is that not getting health care and ending democracy are unnecessary to keep "America great" and White. This is the ultimate and most tragic impact of racism on White Americans.

There is no need for us to plead the case to end racism because it is in all of our better interest. The facts are there, and the reality to staring us in the face. It is time to finally deal with the people in our generation who have used their power and control to keep us divided and less than who we can become. We have the tools to do so. Racism takes away from all of us. It is time to end it and move on.

Discussion Questions

Q1: Who is responsible for racism in the United States?
Q2: Who benefits from racism in the United States? Who is negatively impacted by it?
Q3: What can be done to ameliorate racism in the United States?

References

Jones, James M., and Robert T. Carter. 1996. Racism and White Racial Identity: Merging Realities. In *Impacts of Racism on White Americans*, edited by Benjamin P. Bowser and Raymond G. Hunt, 2nd ed, 1–24. Sage.

Pettigrew, Thomas F. 1981. The Mental Health Project. In *Impacts of Racism on White Americans*, edited by Benjamin P. Bowser and Raymond G. Hunt. Sage.

Spanierman, Lisa, and D. Anthony Clark. 2021. Taking White Racial Emotions Seriously: Revisiting the Costs of Racism to White Americans. In *Impacts of Racism on White Americans in the Age of Trump*, edited by Duke W. Austin and Benjamin P. Bowser, 115–136. Palgrave.

Index

D. W. Austin and B. P. Bowser (eds.), *Impacts of Racism on White Americans In the Age of Trump*, https://doi.org/10.1007/978-3-030-75232-3

Printed by Printforce, United Kingdom